Teacher's Guide to

INVITATION TO PHYSICS

by

Jay M. Pasachoff and Marc L. Kutner

W · W · Norton & Company New York · London

W. W. Norton & Company, Inc.
500 Fifth Avenue
New York, NY 10110

W. W. Norton Company Ltd.
25 New Street Square
London EC4A 3NT U.K.

ISBN 0-393-95164-2

To accompany **Invitation to Physics** by Jay M. Pasachoff and Marc L. Kutner. QC21.2.P38 ISBN 0-393-95152-9

 Prepared by Corona Publishing Co., Inc. Published simultaneously in Canada by George J. McLeod Limited, Toronto. Printed in the United States of America.

Last digit is the print number: 9 8 7 6 5 4 3 2 1

INTRODUCTION

We have prepared this **Teacher's Guide** to assist teachers in the use of our text **Invitation to Physics.** We have thus included the answers to all questions in the text. We have also provided a set of quizzes, one for each section of the text. Answers are supplied. We also include an alternative set of exams, including hour exams and a final.

A **Test Bank** of additional questions suitable for use on exams is available separately. The Test Bank includes chapter-by-chapter sets of multiple-choice questions with answers and additional essay questions. The publisher will supply copies on request.

Because a detailed table of contents, although basically simple and useful, can appear formidable to a student, we have put it in this **Teacher's Guide** instead of the main text. We have also included sample syllabi for the use of **Invitation to Physics** in courses of varying lengths.

The delight that students can feel in taking freshman physics can be a mixture of astonishment at the many wonderful demonstrations that can be performed, excitement at the new ideas of physics that are being developed or discovered, and interest in slides, videotapes, and movies showing phenomena too large or too difficult to demonstrate in the classroom. We thus list some references to sources of this material.

We thank Cathryn Baskin for her excellent work on this book.

We would be very pleased to hear directly from users of our text and its **Teacher's Guide.**

Jay M. Pasachoff
Marc L. Kutner

TABLE OF CONTENTS: TEACHER'S GUIDE

Introduction iii
Table of Contents of **Teacher's Guide** v
Table of Contents of **Invitation to Physics** 1
  Selected List of Essays and Tables 10
Course Syllabi 12
Educational Theory 21
  Piaget 21
  Repechage 24
Apparatus and Laboratory Materials 27
  Labs and Demonstrations 27
  Sources of Equipment 29
  Computers 30
  Holograms and Holography 31
  Funding 31
Organizations and Journals 32
Slides 34
Filmstrips and Films 39
  Filmloops and Filmstrips 39
  Films 48
Answers to Questions at Ends of Chapters 71
Sample Examinations with Answers 111

TABLE OF CONTENTS OF INVITATION TO PHYSICS

CHAPTER 1 THE PHYSICS OF THE UNIVERSE 1–14

1.1 The World of the Small 3
1.2 The Life-Size World 7
1.3 The World of the Large 10
1.4 Why Study Physics? 13

PART I A WORLD OF PARTICLES

Introduction to Part I (Chapters 2–6) 15

CHAPTER 2 PARTICLES IN MOTION 17–39

2.1 Position 18
  2.1a One Dimension 19
  2.1b More Than One Dimension 20
2.2 Motion 21
  2.2a Velocity 21
  2.2b Acceleration 27
2.3 Falling Bodies 29
2.4 Projectile Motion 34
2.5 Relative Motion 36

CHAPTER 3 LAWS OF MOTION 40–57

3.1 Newton's Laws 41
  3.1a First Law--Inertia 42
  3.1b Second Law--Mass 43
  3.1c Third Law--Action/Reaction 45
3.2 Touching and Friction 47
3.3 Statics 47
  3.3a Ropes and Strings 49
  3.3b Springs 50
3.4 Falling Bodies Explained 51
3.5 Accelerating Observers 52
3.6 The Role of Newton's Laws 55

CHAPTER 4 MOMENTUM 58–67

4.1 Momentum 59
4.2 Conservation of Momentum 59
4.3 Rockets 61
4.4 Center of Mass 62

CHAPTER 5 ENERGY 68–84

5.1 Work 69
5.2 Work and Energy 71
5.3 Potential Energy 72
5.4 Conservation of Energy 74
5.5 Energy Conservation and Harmonic Motion 75
5.5a The Pendulum 76
5.5b Springs 77
5.6 Energy Conservation and Collisions 78

CHAPTER 6 ROTATION 85–104

6.1 Circular Motion 85
6.2 Describing Circular Motion 93
6.2a Angular Velocity 93
6.2b Torque 94
6.2c Angular Momentum 95
6.3 Angular Momentum and Extended Objects 96
6.3a Applying Conservation of Angular Momentum 96
6.3b Pulsars 97
6.4 Torques and Extended Objects 99
6.5 Balancing 100

PART II RELATIVITY

Introduction to Part II (Chapters 7–9) 105

CHAPTER 7 GRAVITY 108–128

7.1 The Apple and the Moon 108
7.1a The Motions of the Planets 109
7.1b Kepler's Laws 112
7.1c The Law of Gravitation 114
7.1d The Earth's Gravity 115
7.1e The Strength of Gravity 116
7.2 A Closer Look at Orbits 118
7.3 Systems of Many Bodies 120
7.4 Tides 124

CHAPTER 8 SPECIAL RELATIVITY 129–150

8.1 The Puzzle of Light 130
8.1a The Speed of Light 131
8.1b The Aberration of Starlight 132
8.1c The Michelson–Morley Experiment 134
8.2 The Principle of Relativity 135
8.2a Frames of Reference 135
8.2b The Speed of Light 135
8.3 Problems with Time 136
8.3a Simultaneity 136
8.3b Time Dilation 137
8.3c The Twin Paradox 141
8.4 Spacetime 142
8.5 The Lorentz Transformations 144
8.5a The Lorentz Contraction 145
8.5b The Appearance of Moving Objects 145
8.6 Mass and Energy 146
8.7 Some Concluding Remarks 148

CHAPTER 9 GENERAL RELATIVITY 151–164

9.1 The Principle of Equivalence 152
9.1a Einstein's Elevator 153
9.1b Tests of the Principle of Equivalence 153
9.2 Curved Space 154
9.3 Experimental Tests of General Relativity 158
9.4 Gravity Waves 162

PART III PARTICLES TOGETHER

Introduction to Part III (Chapters 10–12) 165

CHAPTER 10 FLUIDS 167–186

10.1 Describing a Fluid 167
10.2 Fluids at Rest 171
10.3 Buoyancy 172
10.3a Floating 173
10.3b Air Pressure 174
10.4 Fluids in Motion 178
10.4a Flying 180
10.4b Air Flow and Sports 181

CHAPTER 11 TEMPERATURE AND HEAT 187–204

11.1 Temperature 187
11.2 Thermal Expansion 188
11.3 Heat 190
11.4 Heat and Work 192
11.5 Temperature and Energy in an Ideal Gas 192
11.6 The Second Law of Thermodynamics--Are Perfect Engines Possible? 193
11.7 Phase Changes 198
11.8 Cryogenics 201

CHAPTER 12 HOW HEAT GETS AROUND 205–218

12.1 Conduction 206
12.2 Convection 211
12.3 Radiation 213
12.3a Atmospheres 214
12.3b Heating and Cooling 216

PART IV WAVE PHENOMENA

Introduction to Part IV (Chapters 13–18) 219

CHAPTER 13 WAVES 220–235

13.1 Wave Pulses 220
13.1a Strings 221
13.1b Springs 222
13.1c Air 223
13.2 Harmonic Waves 226
13.3 Polarization 230
13.4 The Doppler Effect 231

CHAPTER 14 STANDING WAVES 237–251

14.1 Interference--When Waves Get in Each Other's Way 237
14.2 Reflections--The End of the Line 240
14.3 Standing Waves on a String 242
14.4 Resonance 244

14.5 Standing Sound Waves 246
14.6 The Nature of a Note 248
14.7 Beats 249

CHAPTER 15 LIGHT--THE WAVE SIDE 252-267

15.1 Particle or Wave? 252
15.2 Light Wave Propagation 254
15.3 Interference of Light Waves 256
15.3a Young's Experiment 256
15.3b Interference and Thin Films 258
15.4 Diffraction 259
15.5 Diffraction Gratings and Spectroscopy 260
15.6 Polarization of Light 263
15.7 Doppler Shifts in Light 264

CHAPTER 16 HOLOGRAPHY 268-275

16.1 Lensless Photography 268
16.2 Practical Holography 270
16.3 The Uses of Holography 272
16.4 Holographic Movies 272

CHAPTER 17 OPTICS 276-296

17.1 Reflection and Refraction 276
17.2 Fiber Optics 282
17.3 A Look at Images--Mirror, Mirror on the Wall 282
17.4 Images by Refraction 287
17.5 Colors and Refraction 291

CHAPTER 18 OPTICAL INSTRUMENTS 297-310

18.1 The Simple Camera 297
18.2 The Eye 300
18.2a Focus 300
18.2b Color 301
18.3 Telescopes 302
18.3a The Uses of Telescopes 304
18.3b Types of Telescopes 307
18.4 Microscopes 308

PART V ELECTRICITY AND MAGNETISM

Introduction to Part V (Chapters 19–22) 311

CHAPTER 19 ELECTRICITY AT REST 312–328

19.1 Electric Charge 312
19.2 Coulomb's Law 315
19.3 Electric Fields 317
19.4 Electricity and Energy Conservation 319
19.5 The Oscilloscope 322
19.6 Capacitors--Storage for Charge 323
19.7 High Voltage 324

CHAPTER 20 ELECTRIC CIRCUITS 329–347

20.1 Current in a Conductor 329
20.2 Circuits 334
20.3 Energy in Circuits 337
20.4 Microcircuits 341
  20.4a Integrated Circuits 342
  20.4b The Pocket Calculator 343
20.5 Safety and Electricity 344

CHAPTER 21 MAGNETISM 348–364

21.1 Magnetic Fields and Forces 350
21.2 Magnetic Moments 354
21.3 Where Magnetic Fields Come From 356
21.4 Magnetic Poles 357
21.5 Natural Magnetism .360
21.6 Magnetism in the Solar System 360

CHAPTER 22 CHANGING FIELDS: MOTORS TO MAXWELL 365–378

22.1 Induction 365
22.2 Electricity Goes to Work 368
22.3 Inductance 372
22.4 AC Circuits 373
22.5 Maxwell's Equations and Radiation 373

PART VI QUANTA AND ALL THAT

Introduction to Part VI (Chapters 23–27) 379

CHAPTER 23 QUANTUM THEORY AND ATOMIC STRUCTURE 381–401

23.1 Black-body Radiation--Classical Theory Fails 381
23.2 Photons 383
23.3 Problems in Atomic Structure 386
23.3a Models of the Atom 386
23.3b The Bohr Atom 388
23.4 Matter Waves and Wave Mechanics 389
23.5 Quantum Mechanics 391
23.6 Uncertainty 393
23.7 The Pauli Exclusion Principle and Atomic Structure 395
23.8 The Triumphs of Quantum Mechanics 396
23.9 The Dual Nature of Light and Particles 398

CHAPTER 24 LASERS 402–412

24.1 The Discovery of Masers 402
24.2 The Discovery of Lasers 406
24.3 The Uses of Lasers 408
24.4 Laser Safety 411

CHAPTER 25 SOLID-STATE PHYSICS 413–427

25.1 Structure of Solids 413
25.2 Conductors and Insulators 414
25.3 Semiconductors 417
25.4 Transistors 421
25.5 The MOSFET 424
25.6 Superconductors 424

CHAPTER 26 INSIDE THE NUCLEUS 428–446

26.1 Building a Nucleus 428
26.1a The Building Blocks 428
26.1b The Glue 430
26.1c The Building Code 431
26.2 Mass, Energy, and Stability of Nuclei 433

26.3 Radioactivity 435
26.3a Alpha Decay 436
26.3b Beta Decay 437
26.3c Gamma Decay 439
26.3d Uses of Radioactivity 439
26.4 Nuclear Reactions 440
26.4a Fission 441
26.4b Fusion 443
26.4c Natural, Newly Made, and "Superheavy" Elements 443

CHAPTER 27 ENERGY FOR TOMORROW 447–470

27.1 Fossil Fuels 448
27.1a Petroleum and Natural Gas Use 448
27.1b Coal 451
27.2 Nuclear Energy 453
27.2a Fission Reactors 454
27.2b Breeder Reactors 456
27.2c Side Effects of Radioactivity 457
27.2d Safety of Fission Reactors and of Other Energy Sources 459
27.3 Other Alternatives 462
27.3a Fusion 462
27.3b Solar Energy 465
27.3c Geothermal Energy 467
27.3d Conservation 468
27.4 The Outlook 468

PART VII STRUCTURE OF THE UNIVERSE

Introduction to Part VII (Chapters 28–30) 471

CHAPTER 28 QUARKS: AN ASSORTMENT OF COLORS AND FLAVORS 472–490
28.1 Particles and Forces 472
28.2 Particle Proliferation (The Particle Zoo) 473
28.3 Quarks--Order Is Restored 477
28.4 Color 481
28.5 Charm 482
28.6 Truth and Beauty 484
28.7 Fundamental Particles 486

CHAPTER 29 ASTROPHYSICS 491–514

29.1 The Stars 492
29.1a Classifying Stars 493
29.1b The Neutrino Experiment 494
29.2 Stellar Evolution 495
29.2a The Death of Lightweight Stars 495
29.2b Supernovae 496
29.2c Black Holes 497
29.3 Formation of the Solar System 500
29.3a Other Solar Systems? 500
29.3b Life in the Universe 501
29.4 Galaxies 502
29.5 The Expansion of the Universe 506
29.6 Quasars 507
29.7 The Big Bang 509
29.8 The Future of the Universe 511

CHAPTER 30 PHYSICS NOW 511–520

30.1 Scientific Publication 516
30.2 Grants 517
30.3 Physics Now 518

TABLE OF CONTENTS OF INVITATION TO PHYSICS

SELECTED LIST OF ESSAYS AND TABLES

PART I A WORLD OF PARTICLES (CHAPTERS 1-6)

Scientific Notation 2
Prefixes for Certain Powers of 10 6
Addition and Subtraction of Vectors 20
Components of Vectors 21
Galileo 30
Isaac Newton 41

PART II RELATIVITY (CHAPTERS 7-9)

Gravity and Stars 122
Time Dilation: An Example 140
Astronaut Twins 142
Tachyons 148

PART III PARTICLES TOGETHER (CHAPTERS 10-12)

Demonstrating Bernoulli's Principle 179
Superfluids 184
The Laws of Thermodynamics 201
The Laws of Thermodynamics: An Informal View 202

PART IV WAVE PHENOMENA (CHAPTERS 13-18)

Decibels 226
Gratings 261
Mirages 288
Using a Telescope in Space 304

PART V ELECTRICITY AND MAGNETISM (CHAPTERS 19-22)

The Volt 321
DC and AC Circuits 334
Measuring Magnetic Fields 350
Table 22-1 Regions of the Electromagnetic Spectrum 320

TABLE OF CONTENTS OF INVITATION TO PHYSICS

PART VI QUANTA AND ALL THAT (CHAPTERS 23–27)

Probability 392
Table 23–1 Time Line of the Quantum Revolution 397
Keeping Track of Radioactivity 457
Table 27–1 Average Human Radiation Exposure 458

PART VII STRUCTURE OF THE UNIVERSE (CHAPTERS 28–30)

Symmetry and Conservation Laws 478
Table 28–1 Quarks and Leptons 485
Colliding Beam Electron Accelerator 485
A Free Quark? 486

COURSE SYLLABI

PLAN I: 1 SEMESTER
(36 lectures)

#	Topic	Reading
1	Introduction	Ch 1
2	Motion	2.1–2.4
3	Newton's Laws	3.1–3.2
4	Accelerating Observers	3.4–3.6
5	Momentum	4.1–4.2, 4.4
6	Energy	5.1–5.4
7	Rotation	6.1–6.3
8	Gravity	7.1–7.2
9	Special relativity--history	8.1–8.2
10	Time dilation; Lorentz contraction	8.3, 8.4, 8.6
11	General relativity	9.1–9.3
12	**Hour Exam**	
13	Fluids	Ch 10
14	Heat	11.1, 11.3–11.4, 11.6
15	Heat flow	12.1–12.3
16	Waves	13.1, 13.2, 13.4
17	Standing Waves	14.1, 14.3, 14.5, 14.6
18	Light waves	15.1–15.4
19	Holography	16.1–16.3
20	Optics	17.1, 17.3, 17.4
21	Optical Instruments	18.1–18.3
22	**Hour Exam**	19.1–19.4
23	Electricity	20.1–20.4
24	Circuits	21.1, 21.3, 21.4
25	Magnetism	22.1, 22.2, 22.4, 22.5
26	AC Circuits	
27	Quantum revolution	23.1–23.3
28	Quantum mechanics	23.4–23.9
29	Lasers	24.1–24.3
30	Solids	25.1–25.3, 25.5
31	Nuclei	26.1–26.2
32	Nuclear reactions	26.3–26.4
33	Energy	27.1–27.3
34	Elementary particles	Ch 28
35	Astrophysics	29.2, 29.5, 29.6–29.8
36	Physics now	Ch 30
	Final Exam	

Note: To make room to emphasize modern topics more, omit lectures 4 and 17, and move section 14.1 to lecture 16.

PLAN IIa: ONE TRIMESTER
(30 lectures)

#	Topic	Reading
1	Introduction	Ch 1
2	Motion	2.1–2.4
3	Newton's laws	3.1–3.2
4	Momentum	4.1–4.2, 4.4
5	Energy	5.1–5.4
6	Rotation	6.1–6.3
7	Gravity	7.1–7.2
8	Special relativity	8.1–8.4, 8.6
9	General relativity	9.1–9.3
10	**Hour Exam**	
11	Fluids	Ch 10
12	Heat	11.1, 11.3–11.4, 11.6
13	Heat flow	12.1–12.3
14	Waves	13.1, 13.2, 13.4, 14.1
15	Light waves	15.1–15.4
16	Holography	16.1–16.3
17	Optics	17.1, 17.3, 17.4
18	Optical instruments	18.1–18.3
19	Electricity	19.1–19.4
20	Circuits	20.1–20.4
21	Magnetism	21.1, 21.3, 21.4
22	AC circuits	22.1, 22.2, 22.4, 22.5
23	**Hour Exam**	
24	Quantum revolution	23.1–23.3
25	Quantum mechanics	23.4–23.9
26	Solids	25.1–25.3, 25.5
27	Nuclear physics	26.1–26.4
28	Energy	27.1–27.3
29	Elementary particles	Ch 28
30	Astrophysics	29.2, 29.5, 29.6–29.8
	Final Exam	

PLAN IIb: ONE TRIMESTER
(30 lectures)

Emphasis on Modern Physics

#	Topic	Reading
1	Introduction	Ch 1
2	Motion and Newton's laws	2.2, 2.3, 3.1,3.6
3	Momentum	4.1, 4.2, 4.4
4	Energy	5.1–5.4
5	Rotation	6.1–6.3
6	Gravity	7.1–7.3
7	Special relativity	8.1–8.6
8	General relativity	9.1–9.3
9	Fluids	10.3, 10.4
10	Heat	11.3–11.6, 11.8
11	Heat flow	12.1–12.3
12	**Hour Exam**	
13	Waves	13.1, 13.2, 13.4, 14.1
14	Light	15.1–15.4, 15.7
15	Holography	16.1–16.3
16	Optics	17.1–17.4, 18 1, 18.3
17	Electricity	19.1–19.5
18	Circuits	20.1–20.5
19	Magnetism	21.1, 21.3, 21.6
20	**Hour Exam**	
21	Quantum revolution	22.5, 23.1–23.3
22	Quantum mechanics	23.4–23.9
23	Lasers	24.1–24.3
24	Solids	25.1–25.6
25	Nuclei	26.1–26.2
26	Nuclear reactions	26.3–26.4
27	Energy	Ch 27
28	Elementary particles	Ch 28
29	Astrophysics	29.2, 29.5, 29.6–29.8
30	Physics now	Ch 30
	Final Exam	

More room for expanding modern topics or the treatment of mechanics can be gained if you are willing to eliminate fluids and heat, lessons 9–11 (Chapters 10–12).

PLAN III: 2 SEMESTERS
(36 lectures/semester)

Semester 1

#	Topic	Reading
1	Introduction	Ch 1
2	Describing motion	2.1–2.2
3	Examples of motion	2.3–2.4
4	Relative motion	2.5
5	Newton's laws	3.1
6	Examples of Newton's laws	3.2–3.3
7	Accelerating observers	3.4–3.6
8	Momentum	4.1–4.2
9	Center of mass; ex. of conserv. of momentum	4.3–4.4
10	Work and energy	5.1–5.2
11	Potential energy; conservation of energy	5.3–5.4
12	Simple harmonic motion; collisions	5.5–5.6
13	Circular motion	6.1
14	Angular motion	6.2
15	Conservation of angular momentum	6.3–6.5
16	**Hour Exam**	
17	Law of gravitation	7.1
18	Orbits; tides	7.2–7.4
19	Special relativity—history	8.1–8.2
20	Special relativity and time dilation	8.3
21	Spacetime; Lorentz contraction	8.4–8.7
22	Principle of equivalence	9.1
23	General relativity	9.2–9.4
24	**Hour Exam**	
25	Fluids at rest	10.1–10.3
26	Fluids in motion	10.4
27	Temperature	11.1–11.2
28	Heat, work, first law of thermodynamics	11.3–11.5
29	Second law of thermodynamics	11.6–11.8
30	Conduction	12.1
31	Convection; radiation	12.2–12.3
32	**Hour Exam**	
33	Wave pulses and harmonic waves	13.1–13.2
34	Polarization; Doppler effect	13.3–13.4
35	Interference; standing waves on string	14.1–14.3
36	Standing sound waves; resonance; music	14.4–14.7
	Final Exam	

PLAN III: 2 SEMESTERS
(36 lectures/semester)

Semester 2

#	Topic	Reading
1	Light wave interference	15.1–15.3
2	Diffraction; polarization; Doppler shifts	15.4–15.7
3	Holography	Ch 16
4	Reflection and refraction	17.1–17.2
5	Images by reflection	17.3
6	Images by refraction	17.4–17.5
7	Camera and eye	18.1–18.2
8	Telescopes and microscopes	18.3–18.4
9	**Hour exam**	
10	Electric charge and Coulomb's law	19.1–19.2
11	Electric fields; electric energy	19.3–19.4
12	Oscilloscope; capacitors; high voltage	19.5–19.7
13	Circuits	20.1–20.3
14	Mircocircuits	20.4–20.5
15	Magnetic forces	21.1–21.2
16	Origin of magnetic fields	21.3–21.6
17	Induction	22.1–22.3
18	AC circuits; Maxwell's equations	22.4–22.5
19	**Hour Exam**	
20	Quantum revolution	23.1–23.3
21	Quantum mechanics	23.4–23.6
22	Wave/particle duality; tests of quan. mech.	23.7–23.9
23	Lasers	Ch 24
24	Solids	25.1–25.3
25	Transistors	25.4–25.6
26	Nuclei	26.1–16.2
27	Radioactivity	26.3
28	Fission and fusion	26.4–26.5
29	Energy problem; nuclear energy	27.1–27.2
30	Energy alternatives	27.3–27.4
31	**Hour Exam**	
32	Elementary particles	28.1–28.2
33	Quarks	28.3–28.7
34	Stars	29.1–29.3
35	The universe	29.4–29.8
36	Physics now	Ch 30
	Final Exam	

PLAN IV: 2 SEMESTERS
(24 lectures/semester)

Semester 1

#	Topic	Reading
1	Introduction	Ch 1
2	Describing motion	2.1–2.2
3	Examples of motion	2.3–2.5
4	Newton's laws	3.1–3.3
5	Falling bodies; accelerating observers	3.4–3.6
6	Momentum	Ch 4
7	Work; energy	5.1–5.3
8	Conservation of energy	5.4–5.6
9	Circular motion	6.1–6.2
10	Conservation of angular momentum	6.3–6.5
11	**Hour Exam**	
12	Gravity	Ch 7
13	Special relativity--history	8.1–8.2
14	Time dilation; Lorentz contraction	8.3–8.7
15	General relativity	Ch 9
16	Fluids at rest	10.1–10.3
17	Fluids in motion	10.4
18	Heat; work; first law of thermodynamics	11.1–11.2
19	Second law of thermodynamics	11.3–11.5
20	Conduction; convection; radiation	Ch 12
21	**Hour Exam**	
22	Waves	13.1–13.2
23	Polarization; Doppler effect; interference	13.3–13.4, 14.1
24	Standing waves	14.2–14.7
	Final Exam	

PLAN IV: 2 SEMESTERS
(24 lectures/semester)

Semester 2

#	Topic	Reading
1	Light waves	Ch 15
2	Holography	Ch 16
3	Reflection; refraction; images in reflection	17.1–17.3
4	Images in lenses	17.4–17.5
5	Optical instruments	Ch 18
6	Electric charge, fields	19.1–19.3
7	Electric energy; oscilloscopes	19.4–19.7
8	Circuits	Ch 20
9	Magnetic forces	21.2–21.2
10	Origin of magnetic fields	21.3–21.6
11	Induction and AC circuits	Ch 22
12	**Hour Exam**	
13	Quantum revolution	23.1–23.3
14	Quantum mechanics	23.4–23.9
15	Lasers	Ch 24
16	Solids	Ch 25
17	Nuclei	26.1–26.2
18	Nuclear reactions	26.3–26.5
19	Energy resources	Ch 27
20	**Hour Exam**	
21	Elementary particles	28.1–28.2
22	Quarks	28.3–28.7
23	Stars	29.1–29.3
24	The universe; physics now	29.4–29.8, Ch 30
	Final Exam	

PLAN V: 2 TRIMESTERS
(30 lectures/trimester)

Trimester 1

#	Topic	Reading
1	Introduction	Ch 1
2	Describing motion	2.1–2.2
3	Examples of motion	2.3–2.5
4	Newton's laws	3.1
5	Examples of Newton's laws	3.2–3.3
6	Accelerating observers	3.4–3.6
7	Momentum	Ch 4
8	Work and energy	5.1–5.2
9	Potential energy; conservation of energy	5.3–5.4
10	Simple harmonic action; collisions	5.5–5.6
11	Circular motion	6.1
12	Angular motion	6.2
13	Conservation of angular momentum	6.3–6.5
14	**Hour Exam**	
15	Gravitation	7.1
16	Orbits; tides	7.2–7.4
17	Special relativity--history	8.1–8.2
18	Time dilation; Lorentz contraction	8.3–8.7
19	General relativity	Ch 9
20	Fluids at rest	10.1–10.3
21	Fluids in motion	10.4
22	Heat; work; first law of thermodynamics	11.1–11.5
23	Second law of thermodynamics	11.6–11.8
24	Conduction	12.1
25	Convection; radiation	12.2–12.3
26	**Hour Exam**	
27	Wave pulses and harmonic waves	13.1–13.2
28	Polarization; Doppler shifts	13.3–13.4
29	Interference; standing waves on string	14.1–14.3
30	Standing sound waves; resonance; music	14.4–14.7
	Final Exam	

PLAN V: 2 TRIMESTERS
(30 lectures/trimester)

Trimester 2

#	Topic	Reading
1	Light wave interference	15.1–15.3
2	Diffraction; polarization; Doppler shifts	15.4–15.7
3	Holography	Ch 16
4	Reflection	17.1–17.3
5	Refraction	17.4–17.5
6	Optical instruments	Ch 18
7	Electric charge; fields	19.1–19.3
8	Electric energy	19.4–19.7
9	Circuits	20.1–20.3
10	Microcircuits	20.4–20.5
11	Magnetic forces	21.1–21.2
12	Origin of magnetic fields	21.3–21.6
13	Introduction; AC circuits	22.1–22.5
14	**Hour Exam**	
15	Quantum revolution	23.1–23.3
16	Quantum mechanics	23.4–23.6
17	Wave/particle duality; tests of quantum mech.	23.7–23.9
18	Lasers	Ch 24
19	Solids	25.1–25.3
20	Transistors	25.4–25.6
21	Nuclei	26.1–26.2
22	Nuclear reactions	26.3–26.5
23	Energy problem; nuclear energy	27.1–27.2
24	Energy alternatives	27.3–27.4
25	**Hour Exam**	
26	Elementary particles	28.1–28.2
27	Quarks	28.3–28.7
28	Stars	29.1–29.3
29	The universe	29.4–29.8
30	Physics now	Ch 30
	Final Exam	

PIAGET

The psychological theories of Jean Piaget have been much discussed in educational circles recently, and are being applied to the teaching of physics and other sciences. Workshops have been devoted to the application of the theories of Piaget.

Often, however, the theories are applied in a simplified form. Sometimes the last two of Piaget's four stages are the only ones discussed, and from the conclusion that most students reason at only the third, "concrete," level instead of the fourth, "formal," level, it is suggested that teaching methods or the type of material presented should be modified.

It is not so often realized by physicists that the views of Piaget are controversial in the psychological community, and that other viable theories are in conflict with Piaget's. D. C. Phillips of Stanford University and Mavis E. Kelly of Monash University, in their article, "Hierarchical Theories of Development in Education and Psychology" (**Harvard Educational Review**, 45, No. 3, August 1975, pp. 351-375), conclude that their examination of "the much-touted hierarchical theories of development" has shown that "their underlying asuumptions have not been adequately examined." Ralph A. Goodwin, a physicist at the U.S. Naval Academy, expresses skepticism in "Talk and Chalk" (**Physics Teacher**, September 1978, pp. 367-371; he also discusses the Keller Plan there).

The validity of the Piagetian theory has been examined by H. Daniel Cohen, Donald F. Hillman, and Russell M. Agne, all of the University of Vermont. Prof. Cohen is from the Department of Physics and Profs. Hillman and Agne are from the College of Education and Social Services. They chose 195 students randomly from introductory physics courses and tested them with selected Piagetian tasks. "Though 58% of the sample were rated at the level of concrete operations, the Piagetian level was found to be only weakly correlated with both level of course and with final course grade." They conclude, "that there exist different developmental levels of cognitive ability among college students as determined by Piaget's measures, is clearly demonstrated," but "that these developmental differences warrant new approaches to curriculum, teaching methods and evaluation procedures, does not appear to be supported by our results. Our data show little correlation between Piagetian level and success in physics courses. In particular, the most advanced course, hence likely the most formal, showed no correlation between Piagetian level and final course grade."

Piaget and his work are treated in a biography, **Jean Piaget**, by Margaret A. Boden (New York: Viking Press, Modern Masters, 1980). Boden says, "Two general (and connected) criticisms may be made, each of which should be borne in mind. . .because they apply also to his accounts of behavior in older children. First, he tends to overestimate the unity and distinctness of each stage. . . . Second, Piaget tends to underestimate the complexity of children's cognitive achievements." Piaget "disagrees with education-alists such as J. S. Bruner, who claims that 'any idea or problem or body of knowledge can be presented in a form simple enough so that any particular learner can understand it in a recognizable form.'"

John Leonard's review in **The New York Times** for January 25, 1980, says that Boden, "reader in philosophy and psychology at the University of Sussex, has doubts about Piaget's prose style, his theory, his methodology and his math. . . ." After a brief discussion of the theory, he asks, "Even if it isn't clear, is it true? I don't know. Obviously, we have trouble thinking <u>about</u> thinking in words that make sense. What I call 'spontaneous learning' is what Piaget and Dr. Boden call 'equilibratory cognitive improvements.'

"Dr. Boden is firm when clinical experience and scientific experiment contradict Piaget's assertions. She reminds us that he relies too much on anecdotes. She is impatient with his indifference to social factors--mother and baby, for instance, plus language--in cognitive development. She doubts whether his stages are 'invariable,' and even wonders if there are any stages at all. And she has interesting things to say about his pretensions as a biologist (he wrote his doctoral thesis on molluscs) and philosopher (a little Kant and a lot of mush, it seems to me).

"On the whole, however, she approves, and I suppose the rest of us must too. . . . The vagueness wouldn't bother me so much, were it not for the fact that the vagueness and the anecdotes are translated into educational theory and plunked right down in American schools, where reading is no longer very important because learning to read somehow violates the 'rhythm' of a child's inevitable development. I'm not sure his three white upper-middle-class Swiss children, now immortalized in 60 books, have much to say to the hundreds of thousands of nonwhite, non-upper-middle-class non-Swiss children now neglected by our schools."

Several studies of Piaget's theories and their application to teaching are now going on. All current reviews of Piaget's work agree that the Piagetian stages are not as discrete as he had thought. Insofar as professors give more attention to teaching in ways that students can understand, then the attention to the work of Piaget by physicists is good. Insofar as a superficial application of Piaget's theories leads to material being omitted entirely, then we must be very cautious.

REPECHAGE

I have several times tried out a new system of student-teacher involvement called **repechage** (re'pe-chaj). It was suggested by Martin S. Weinhous and William R. Webber ("Repechage: A Second Chance at Learning," **American Journal of Physics**, 43, No. 9, 1975, pp. 831-832). The system is described in the material that follows, which can be distributed to each student at the beginning of the semester.

Basically, the method involves retesting the student after each hour exam on the material that he or she missed and **only the material that was missed** on the exam. The advantage for the student is that up to half credit can be restored; from the professor's point of view, the main advantage is the ability to isolate the particular points that have proved difficult to students.

The system requires some additional work on the part of the professor, and there are indeed times (usually right before the final deadlines) when you would probably wish that you had never gotten into it. But the advantages of providing individual contact with each student are many, and the repechage system makes a much higher percentage of that contact functional and instructive, as opposed to the Keller plan, in which a lot of time is taken up in routine matters. The repechage time allotment can work out to about 5 minutes per student per hour exam. It does usually require the professor because a teaching assistant does not necessarily have either the breadth of knowledge or the authority to deny automatic restoration of credit if a student learns only the particular bit of information missed but not the whole subject.

The advantages of the system include, for the professor, a much greater understanding of where students are going wrong and how the course is going. It also leads to a personal knowledge of each student that is not only interesting and pleasant on the part of the professor but also very desirable and appreciated on the parts of the students. In places where student ratings play a role in the promotion of instructors, the increase in personal interest that students perceive and report on questionnaires can be useful and important to instructors. Almost all students rate the repechage system highly, with only a very small residuum protesting that they don't want to study any more once the exam is over.

The following page may be distributed to students.

REPECHAGE

On the hour exam, we will use a system that enables a student to master the material and be retested. The process is called "repechage," after the procedure in crew where the boats that do not win in their respective heats go into "repechage" runoffs, the winners of which are included in the finals. The system emphasizes the fact that the exams are meant to be learning experiences rather than merely testing; your responsibility to have the material in the course mastered does not end when you leave the exam room.

Tentative letter grades will be assigned when the bluebooks are returned by following the grade curve that is listed elsewhere; these grades are only preliminary. In the two weeks following the return of each hour exam, I will schedule extra office hours in addition to my regular office hours. Each student, having reviewed his or her bluebook at home in order to understand why credit was lost and having determined what the correct answers are, must see me during those office hours. It is not sufficient to know only the correct answers to the specific questions on the exam; each student must be able to show that he or she now understands the subject about which he or she answered incorrectly. If the student shows in the repechage that he or she has now mastered the material, half the points previously subtracted will be awarded. Thus if the original score was 60%, it is possible to come out of the repechage with a grade of 80%.

At the end of the repechage period, a new curve will be computed for the course, following the same percentage distribution of letter grades that was used before. Students who have mastered in the repechage process all the material they previously got wrong will wind up with the same letter grade as their preliminary letter grade, or better. Students who came unprepared to the repechage, or who did not come to the repechage, will have slipped behind the others and will wind up with worse letter grades than their preliminary grades. If everyone masters all the material and gets half-credit back in the repechage, then the relative standings of the members of the class will not change and the letter grades will be unchanged. Every student--even those with an A+--is expected to participate.

LABS AND DEMONSTRATIONS

The basic books on physics demonstrations are:

1. Meiners, Harry F., ed. and Robert Resnick, project co-chairman, **Physics Demonstration Experiments**, 2 vol. Sponsored by the AAPT. (New York: The Ronald Press for the AAPT, 1970). The 2-volume set is also available for $43.95 from John Wiley & Sons, Inc., 605 Third Ave., New York, NY 10016. (Reviewed in **American Journal of Physics** 40, 641, 1972; **Physics Today** 24, 50, 1971; **Physics Teacher 10**, 482, 1971.)

2. Sutton, Richard Manliffe, **Demonstration Experiments in Physics** (New York: McGraw Hill, 1938). Out of print, but nonetheless still in general use. (Reviewed in **American Journal of Physics 3**, 85, 1935.)

Additional valuable books are:

1. Frier, George D., and F. J. Anderson, **A Demonstration Handbook for Physics** (Minneapolis: Frier and Anderson, 1972). Paperbound, 211 pp. $5.00. Out of print, but a revision will be published in 1981, probably by the American Institute of Physics. Contact the AIP or Prof. George D. Frier, School of Physics and Astronomy, University of Minnesota, Minneapolis, MI 55455. (1972 edition reviewed in **Physics Teacher 11**, 252, 1973.)

2. Hilton, Wallace A., **Physics Demonstration Experiments at William Jewell College.** The book is currently out of print, but the Apparatus Committee of the AAPT is considering printing it.

3. Davis, John A., "Physics Lecture Demonstrations: An Annotated Bibliography, Part I," **Physics Teacher 12**, 523 (1974).

4. Kutasov, Shoma, **Physics Demonstrations** (Los Angeles: Penn Books, 954 7th St., Santa Monica, CA 90403, 1978). 302 pp., $12.85, paperbound. Contact Penn Books or Dr. Kutasov,

Dept. of Physics, Occidental College, 1600 Campus Rd., Los Angeles, CA 90041. Experiments from Soviet secondary schools. (Reviewed in **Physics Teacher**, November 1979, pp. 542–543.) Includes a semiconductor demonstration set, consisting of six panels: solar cell, thermistor, two diodes, transistor, photoresistor, and high sensitivity relay. $275.

Other books and sources are listed in **Resource Letter PhD-1: Physics Demonstrations**, prepared as a project of the AAPT Committee on Apparatus for Educational Institutions. It appeared in the **American Journal of Physics** 47 (10), October 1979, pp. 835–840. Copies are available if you mail 30 cents per copy (not in stamps) together with a stamped and self-addressed envelope to:

Executive Office
American Association of Physics Teachers
Graduate Physics Building
SUNY at Stony Brook
Stony Brook, NY 11794.

(The multi-number code at the bottom of the first page of the article shows that you are supposed to pay $0.50 if you Xerox or otherwise copy it.)

APPARATUS AND LABORATORY MATERIALS

SOURCES OF EQUIPMENT

Ealing
Physics Teaching Equipment
Pleasant St.
South Natick, MA 01760

Edmund Scientific
330 Edscorp Building
Barrington, NJ 08007

Fisher Scientific Co.
Educational Materials Division
4901 West Lemoyne Ave.
Chicago, IL 60651

Damon/Instructional Systems Division
80 Wilson Way
Westwood, MA 02090

Write for equipment for use with PSSC Physics, published by D.C. Heath and Co.

Klinger Educational Products Corp.
83-40 Parsons Blvd.
Jamaica, NY 11432

Send for Bulletin 114 for information on the Blackboard Optics Kit, an assembly of large plastic lenses and mirrors that may be secured to a classroom blackboard.

Metrologic
143 Harding Ave.
Bellmawr, NJ 08031

PASCO Scientific
1933 Republic Ave.
San Leandro, CA 94577

Tel-Atomic, Inc.
P.O. Box 3444
Charlotte, NC 28203

Have available a low-cost x-ray spectrometer, along with lab manual with experiments.

COMPUTERS

Apple Computers
10260 Bandley Drive
Cupertino, CA 95014
Write for an educational information packet.

Commodore Corp.
PET Systems Division
3330 Scott Blvd.
Santa Clara, CA 95051

Control Data Corp.
Education Industry Marketing
HQW09A, Box O
Minneapolis, MN 55440
Write for information on their computer-based physics programs.

North Star Computers, Inc.
1440 Fourth St.
Berkeley, CA 94710

Ohio Scientific
1333 S. Chillicothe Rd.
Aurora, OH 44202

Radio Shack
Division of Tandy Corp.
Dept. CE-081
1400 One Tandy Center
Fort Worth, TX 76102
Write for catalogue and information on title funding.

ADDITIONAL INFORMATION:

Microcomputers in the Classroom, a reprint available from The Physics Teacher, AAPT, Graduate Physics Bldg., SUNY at Stony Brook, Stony Brook, NY 11794. Included are 5 articles about teaching with microcomputers.

Those interested in microcomputers should look at:

Byte (Monthly)
Subscription Dept., PO Box 590, Martinsville, NJ 08836
800/258-5485; 1 year $18, 2 years $32, 3 years $46.

onComputing (Quarterly guide to personal computing)
Subscription Dept., PO Box 307, Martinsville, NJ 08836
1 year $8.50

"Exploring Ballistics with Your Computer," by Robert W. Jenks, in Byte, September 1980, pp. 270-280. See this article for examples on how microcomputers can be used for realistic exercises in which forces like air resistance are not neglected.

HOLOGRAMS AND HOLOGRAPHY

Multiplex
454 Shotwell St.
San Francisco, CA 94110

Holographic Road Show
Larry Kirkpatrick and Mac Rugheimer
Montana State University
Bozeman, MT 59717

A traveling exhibit that consists of an initial viewing session, a short lecture, and a second viewing session. (See Physics Teacher, January 1979, pp. 25–31.)

Workshops in holography are held at Lake Forest College. Each five-day workshop provides experience in the construction and showing of holograms, and discussion includes theory, development, and applications of holography. For further information, write:

T. H. Jeong, Director
Holography Workshops
Lake Forest College
Lake Forest, IL 60045

* * * * *

FUNDING

Note that the Instructional Science Equipment Program (ISEP) of the National Science Foundation, Washington, DC 20550, has provided matching funds for teaching equipment. Check with the NSF to see if the program is still in operation in any given year.

THE AAPT

The American Association of Physics Teachers
Graduate Physics Building
SUNY at Stony Brook
Stony Brook, NY 11794

The AAPT is the principal organization of physics teachers in the United States. They publish:

The American Journal of Physics, a major refereed journal containing technical articles relating to the teaching of physics. The AJP also contains book and film reviews and apparatus notes. Manuscripts for publication should be submitted to:

John S. Rigden, Editor
American Journal of Physics
240 Benton Hall
University of Missouri at St. Louis
St. Louis, MO 63121

For preparation of manuscripts see the inside cover of the January 1980 issue.

Physics Teacher, which is read (judging from articles and letters) more by high school or junior college teachers than by college teachers, but which contains much of interest to everybody. **Physics Teacher** includes articles, apparatus notes, and such monthly features as, "Would you believe...?," "String and sticky tape experiments," "How things work," "Good reading from other journals," "Book reviews," and "Media Matters" (notes of errors in books).

The AAPT publishes a catalogue, **AAPT Products**, listing publications, resource letters, reprint books, super-8 film cartridges, slides, modules, and other services available from them.

As of 1980, dues were:

Regular Membership (choice of one journal)	$27.00
Regular Membership (including both journals)	$42.00

Members also receive the **AAPT Announcer**, which contains the programs for the twice-yearly meetings and also contains advertisements and listings of new materials available.

The AAPT usually has two yearly meetings, a winter meeting

that is joint held with the American Physical Society, and a meeting toward the end of June.

Another journal of widespread interest among teachers of physics is:

Physics Today
American Institute of Physics
335 East 45 St.
New York, NY 10017.

Physics Today contains many review articles. It also has comments on current developments in physics ("Search and Discovery") and on physics research and funding ("State and Society").

* * * * *

Society of Physics Students

Students may be interested in joining the Society of Physics Students. Members of the SPS may purchase, for personal use only, the 1980-1981 **Graduate Programs in Physics, Astronomy and Related Fields** at the student rate of $7.50 prepaid. Members of the SPS also receive the **SPS Newsletter.** To join, students should write to:

Membership Department
American Institute of Physics
335 East 45 St.
New York, NY 10017

Dues: Members, $7. Joint student memberships are available with APS, OSA, ASA, AAPT, AA, and AVS.

SLIDES

SOURCES OF SLIDES

The AAPT makes available at least 44 sets of slides covering many topics of physics. Each set comes with a Teacher's Guide. The sets are:

SL-1 Apparatus at the Science Museum of London
SL-2 Archimedes' Principle
SL-3 Atomic Spectra and the Zeeman Effect
SL-4 Atmospheric Phenomena
SL-5 Atmospheric Phenomena II
SL-6 Bentley's Snow Flakes and Ice Crystals
SL-7 Bursting Projectiles
SL-8 Dislocations
SL-9 Electron Distributions in the Hydrogen Atom
SL-10 Entropy and Dot Patterns
SL-11 Entropy of Mixing
SL-12 Equilibrium Examples
SL-13 Frost Patterns and Slushcicles
SL-14 Kinematics Graphs
SL-15 Lights on Campus
SL-16 Mars--The Views from Viking
SL-17 Moire Patterns and Reflection Streaks
SL-18 Multicomponented Measures
SL-19 Nineteenth Century Physics Apparatus
SL-20 Nineteenth Century Textbook Illustrations
SL-21 Non-Commutivity of Angular Displacements
SL-22 Optical Effects at Home
SL-23 PASCO's Drawings of Famous Physicists
SL-24 Photoelectric Plastic Articles
SL-25 Photographs and Drawings of Well-Known Physicists
SL-26 Physics Buttons
SL-27 Pinhole Camera Effects
SL-28 Plates, Bars, Membranes, and Beats
SL-29 Probabilities and Errors, Part 1
SL-30 Probabilities and Errors, Part 2
SL-31 Probabilities and Errors, Part 3
SL-32 Schematic View of the Earth's Magnetosphere
SL-33 Stars and Telescopes
SL-34 Statistics of Splattering Drops
SL-35 Stop-Action Photography
SL-36 Thermal Patterns in Snow
SL-37 Vector Addition

SLIDES

SL-40 Mars: More Views from Viking
SL-41 Total Internal Reflection and Multimode Optical Fiber
SL-42 The Search for Electromagnetic Induction
SL-43 Jupiter
SL-44 The Moons of Jupiter

For further ordering information, write:

AAPT Executive Office
Publications Department
Graduate Physics Building
SUNY at Stony Brook
Stony Brook, NY 11794

* * * * *

Physics Transparencies. Four packets of 6 overhead transparencies each for illustrating ray tracing are available from:

American Animation, Inc.
6 East Spruce Street
P.O. Box 335
Moorestown, NJ 08057

Each packet includes 6 transparencies. The cost is $47.50 per packet. They are:

API-1 Basic Light Phenomena
API-2 The Refraction of Light and Internal Reflection
API-11 Images in Plane and Spherical Mirrors
API-112 Thin Lenses

A spinning Polaroid mounted on your overhead projector makes it look as though the light rays are moving. (Review: **Physics Teacher**, December 1979, p. 614.)

* * * * *

Several series of slide sets are available from:

Projected Learning Programs, Inc.
P.O. Box 11857
Reno, NV 89510

Each slide program is accompanied by a teacher's guide.

SLIDES

Among the slide programs available are:

General Physics (each set costs $13.95):

- Force
- Velocity and Acceleration
- Work, Energy and Power
- Newton's Laws of Motion
- Circulatory and Oscillatory Motion
- Machines
- Current Electricity
- Pressure
- Density and Flotation
- Gases and Their Laws
- Mass, Length and Time
- Waves and Wavefronts
- Magnetic Effects of Currents
- Heat and Temperature
- Resistance and Voltage Measurement
- Ammeters and Voltmeters
- Electrons: Properties and Effects
- Static Electricity

Advanced Physics (each sets costs $13.95):

- The Forces of Nature
- The Hydrogen Atom, Energy Levels
- The Sodium Atom, Spectrum and Quantum Numbers
- An Introduction to Lasers
- Gas Lasers
- Solid-State Lasers

Electronics (each set costs $13.95):

- Electrical Charge
- The Fundamental Particles
- Atoms and Energy
- Free Electrons
- Electric Current
- Conductors and Resistors
- Capacitors and Inductors
- Thermionic Values
- Semiconductor Devices

SLIDES

Nuclear Physics (slides or filmstrips)

These programs were developed at the Educational and Training Centre at Harwell, the United Kingdom's atomic research center (slides: $42.95; filmstrips: $29.95).

* * * * *

Other sources of slides include:

MMF Corp.
3016 Cresmont Ave.
Baltimore, MD 21211
Catalogue of slides, prints, and books ($1).

Norton Scientific
P.O. Box 4294
Tucson, AZ 85717
Catalogue 105. Scientific Slide and Auxiliary Projector Catalogue.

Edmund Scientific Co.
300 Edscorp Building
Barrington, NJ 08007
(tel. 609/547 3488)

Science Graphics
P.O. Box 17871
Tucson, AZ 85731

* * * * *

Astronomical slides are available from:

Astronomical Society of the Pacific
1244 Noriega St.
San Francisco, CA 94122

Hansen Planetarium
Space Science Library and Museum
15 South State St.
Salt Lake City, UT 84111

SLIDES

Palomar Observatory
Caltech Bookstore
Pasadena, CA 91126

Kitt Peak National Observatory
P.O. Box 26732
Tucson, AZ 85762
(send $1 for the catalogue)

Lick Observatory
University of California at Santa Cruz
Santa Cruz, CA 95064

A set of 40 color slides and 250 black-and-white overhead transparencies accompanies adoptions of Contemporary Astronomy, 2nd edition, by Jay M. Pasachoff (Saunders College/HRW, 1981); University Astronomy by Jay M. Pasachoff and Marc L. Kutner (W. B. Saunders Co., 1978), and other alternative versions of these texts.

FILMSTRIPS AND FILM LOOPS

HOLT/EALING

Holt/Ealing has an extensive selection of film loops on a wide variety of topics in physics. The film loops may be purchased directly from Holt, Rinehart & Winston, and are available for $24.95. For a catalogue, write:

Holt, Rinehart & Winston
Media Services Department
383 Madison Ave.
New York, NY 10017

The catalogue also lists the addresses of local distributors of the film loops.

KALMIA FILM LOOPS

The following series of film loops can be purchased not only from Holt/Ealing but also from Kalmia, Dept. P5, Concord, MA 01742. They are all Super 8 mm, silent, Technicolor Cartridge format of approximately 3–4 min. each. Purchase price for any individual film is $28.95 each, and $2.00 less per film when ordered in complete sets.

Drops and Splashes. Color, $28.95. Harold G. Edgerton and Tsuneyoshi Uyemura.

Electromagnetism Series. 12 films, color, $28.95 each, $323.400 per set. Richard B. Adler. **The Concept of Changing Flux; Faraday's Law of Induction; The Field as a Vector; The Field from a Steady Current; Field: The Force on a Current; Field vs. Current; Field vs. Distance; The Magnetic Field; Meters; Monopoles and Dipoles; Motors; Uniform and Non-Uniform Fields.**

Mechanics on an Air Table Series. 9 films, color, $28.95 each, $242.55 per set. David Kutliroff. **The Center of Mass; Circular Motion; Collisions in Two Dimensions; Distance, Time and Speed; Dynamics of**

Circular Motion; Dynamics of Pendulums; One-Dimensional Acceleration; Simple Harmonic Motion; Trajectories.

Wave Motion in a Ripple Tank Series. 14 films, b/w, $28.95 each, $377.30 per set. Produced by Educational Services, Inc. **Barrier Penetration by Waves; Bragg Reflection of Waves; Circular Wave Reflection from Various Barriers; Diffraction and Scattering Around Obstacles; Doppler Effect; Effect of Phase Differences Between Sources; Formation of Shock Waves; Interference of Waves; Multiple Slit Diffraction; Reflection of Waves from Concave Barriers; Refraction of Waves; Single Slit Diffraction; Straight Wave Reflection from Straight Barriers; Superposition of Pulses.**

Explorations in Space and Time Series. 12 computer-animated film loops, color, $28.95 each, $244.00 per set. Produced by M. L. Meeks. Also distributed by Houghton-Mifflin Co., Film Department, Two Park St., Boston, MA 02107. **The Star System Xi Ursae Majoris; Sirius and the White Dwarf; Algol, the Demon Star; The Motions of Stars, Part 1; The Motion of Stars, Part 2; Star Clusters, Part 1; Star Clusters, Part 2; The Motions of Attracting Bodies, Part 1; The Motions of Attracting Bodies, Part 2; The Doppler Effect in Sound and Light; Planetary Motion and Kepler's Laws, Part 1; Planetary Motions and Kepler's Laws, Part 2.**

Kalmia also now distributes many other physics films (including those formerly listed in the Ealing catalogue). Write to Kalmia for their latest catalogue.

EDUCATION DEVELOPMENT CENTER

The Education Development Center produced the **Quantum Physics Series**, a series of 9 computer-animated film loops described below under the chapter on Quantum Theory. Also available from EDC is a tenth film on quantum physics, **Individual Events in One-Dimensional**

Scattering (H. M. Schey, J. L. Schwartz, and R. Whiddon). Distributed in 16mm format by EDC Distribution Center, 55 Chapel St., Newton, MA 02160. Contact the Film Librarian, EDC Distribution Center for distributors in Super 8mm Cartridge format, and see listing below for other distributors. EDC Price (16mm):

16mm Rental (10 titles, composite reel)	$35
16mm Purchase (10 titles, individually)	$30
16mm Purchase (10 titles, series)	$270

For a catalogue listing other film loops and films, write to the EDC Distribution Center.

EXPERIMENTS IN PHYSICS SERIES

Produced by Interopa Film GmbH, Berlin. Distributed by Walter de Gruyter Inc., 3 Westchester Plaza, Elmsford, NY 10523. These are short Super 8 mm, color, Kodak cartridge, silent films. They include from 2 to 10 slides (35 mm) and a text booklet all packaged in a book-type plastic box. Price: $59 purchase each. (Additional films from this series are also listed under the individual chapter headings.)

Electromagnetic Oscillator. I. Free Oscillations. 3.3 min. RLC Circuit via oscilloscope.

Electromagnetic Oscillator. II. Forced Oscillations. 7.4 min. Current, voltage, and phase of weakly damped circuit via double-beam oscilloscope.

Fluid Helium. 5.2 min. Demonstrates superconductivity, trapped flux, and Meissner-Oschsenfeld effect.

Hypersonic Stream. 5 min. Compression waves and Mach cones via Schlieren effect.

Inclined Plate in a Flow. 2.4 min. Demonstrations of vortex free flow, real flow with vortex formation, and torque on an inclined plate.

WOMEN IN SCIENCE by Dinah L. Moche

Includes an audio cassette, slides, and articles about six women with careers in science and engineering. It is available for $30 prepaid, or $31 if billed, from

AAPT Publications Dept.
Graduate Physics Bldg.
SUNY at Stony Brook
Stony Brook, NY 11794

The scientists included in this series are:

Virginia Trimble, Associate Professor of Physics at the University of California, Irvine, and Associate Professor of Physics and Astronomy at the University of Maryland, College Park;

Diana McSherry, a research biophysicist and Executive Vice President of Digisonics Corporation;

Gwendolyn Albert, an environmental engineer with the Southwestern Division of the U.S. Army Corps of Engineers;

Chien Shiung Wu, Professor of Physics at Columbia University;

Betsy Ancker-Johnson, Associate Director of Argonne National Laboratory in charge of physical research;

Carolyn Leach, Chief of the Biomedical Laboratories Branch at the NASA-Johnson Space Center, Houston, Texas.

ADDITIONAL FILM LOOPS AND FILMSTRIPS
(listed in order of chapters in **Invitation to Physics**)

CHAPTERS 2-5 MECHANICS

Experiments in Physics--Mechanics Series.
Edited by Rudolph Weber. Scientific direction by Hans Boersch, assisted by Lothar Schmidt and Walter Ludwig. See ordering information above.

Deals primarily with oscillations and wave motion with two film loops on collisions and two on angular momentum. (Review: American Journal of Physics, Vol. 45, No. 2, February 1977, pp. 230-231.)

CHAPTER 7 GRAVITY

Planetary Motion: Early Ideas.
Universal Gravitation.
Gravitational Forces and Fields.
This filmstrip-cassette series is from **Physics: Fundamental Concepts.** Produced by Lester G. Paldy and Jacqueline D. Spears, Prentice-Hall Media, New York, 1978. Price: $35 each.

Planetary Motion: Early Ideas gives a historical treatment of astronomical theories from the early Greek philosophers to Kepler. **Universal Gravitation** gives a historical presentation of ideas concerning gravity prior to Newton and summarizes ideas that contributed to Newton's reasoning. **Gravitational Forces and Fields** explains gravitational forces using Newton's concept of gravitational force. (Review: Physics Teacher, February 1980, pp. 154-155.)

CHAPTERS 8 AND 9 RELATIVITY

General Relativity: A Theory of Gravitation.
From the series, **Physics: Fundamental Concepts.** Produced by Lester G. Paldy and Jacqueline D. Spears. Prentice-Hall Media, New York, 1978. Price: $35.

Describes Einstein's general theory of relativity as the first inclusive theory of gravitation. (Review: Physics Teacher, February 1980, p. 154.)

Relativity Series.

A series of 8 computer-animated film loops, color (also available in 16mm). Produced by Robert Erhlich. Distributed by Kalmia. Purchase Price: $28.95 each, $400.00 per set. See ordering information above.

The Speed of Projectiles, Sound, and Light; Simultaneity Is Relative; The Michelson-Morley Experiment; Length Contraction; Time Dilation; The Doppler Effect and the Twin Paradox; World Lines; Coordinate Transformation.

CHAPTER 10 FLUIDS

Potential Flow.
Reduction of Flow Cross Section.
Laminar and Turbulent Flow in a Tube.
Objects in Real Flow.
Inclined Plate in Flow.
Distribution of Pressure in Real Flow.
Forces and Torques in a Real Flow.
Magnus Effect.
Vortices.
Initial Vortex.
Hypersonic Stream.

From **Experiments in Physics: Aeromechanics, Hydromechanics and Acoustics Series.** See ordering information above.

(Review: American Journal of Physics, Vol. 45, No. 6, June 1977.)

CHAPTER 11 TEMPERATURE AND HEAT
CHAPTER 12 HOW HEAT GETS AROUND

Adiabatic Expansion I. Clement-Desormes, CO_2 dry ice.
Adiabatic Expansion II. N_2 liquefaction.
Brownian Motion.

From **Experiments in Physics: Thermodynamics, Optics, and Electricity Series.** See ordering information above.

(Review: American Journal of Physics, Vol. 45, No. 10, October 1977, pp. 1014-1015.)

Kinetic Theory by Computer Animation Series.
11 computer-animated film loops, color (also available in 16mm). Produced by John T. Fitch, J. L. Kinsey, and Sherrill F. Martin. Distributed by Kalmia. Purchase Price: $28.95 each, $296.45 per set. See ordering information above.

Pressure, Volume and Boyle's Law; Temperature, Energy and Thermal Equilibrium; Heating, Cooling and Charles' Law; Avogadro's Principle; Ideal Gas Law; Dalton's Law; Gravitational Distribution; Maxwell-Boltzmann Distribution; Brownian Motion and Random Walk; Graham's Law; Deviations from Ideal Gas Law.

CHAPTER 13 WAVES
CHAPTER 14 STANDING WAVES

Standing Acoustic Waves.
Beats of Acoustic Oscillations.
Two film loops from the **Experiments in Physics: Aeromechanics, Hydromechanics and Acoustics Series.** Edited by Rudolph Weber. See ordering information above.

(Review: American Journal of Physics, Vol. 45, No. 6, June 1977, p. 596.)

CHAPTER 17 OPTICS
CHAPTER 18 OPTICAL INSTRUMENTS

Part 1/Reflection of Light: Plane Mirrors and Optical Fibers.
Part 2/Reflection of Light: Curved Mirrors.
Part 3/Refraction of Light: Lenses.
Part 4/Diffraction and Interference.
Four filmstrip-cassettes from the Optics Program, **Physics: Fundamental Concepts.** Lester G. Paldy, Jacqueline D. Spears. Prentice-Hall Media, New York, 1978. Price: $35 each.

(Review: Physics Teacher, February 1980, pp. 155, 158.)

Lens Aberrations Parts 1 and 2.
Image Formation in the Microscope.
Two film loops from the **Experiments in Physics:**

Aeromechanics, Hydromechanics and Acoustics Series. Edited by Rudolph Weber. See ordering information above.

(Review: American Journal of Physics, Vol. 45, No. 10, October 1977, pp. 1014-1015.)

CHAPTERS 19-20 ELECTRICITY

Capacitor I. Voltage and force.
Capacitor II. Dipoles and dielectrics.
R, C, L in an Arc Circuit.
Electromagnetic Oscillator I. Free oscillations.
Electromagnetic Oscillator II. Forced oscillations.

Film loops from the **Experiments in Physics: Aeromechanics, Hydromechanics and Acoustics Series.** See ordering information above.

(Review: American Journal of Physics, Vol. 45, No. 10, October 1977, pp. 1014-1015.)

CHAPTER 23 QUANTUM THEORY AND ATOMIC STRUCTURE

Part 5/Modern Concepts of Light.

Filmstrip-cassette from the Optics Program, **Physics: Fundamental Concepts.** Lester G. Paldy, Jacqueline D. Spears. Prentice-Hall Media, New York, 1978. Price: $35.

Considers the controversy between the particle and wave theories of light. (Review: Physics Teacher, February 1980, p. 158.)

Quantum Physics Series.

Free Wave Packets; Packets in a Perfect Crystal; Packets Incident on a Crystal; Particle in a Box; Scattering from Impurities; Scattering in One Dimension, I: Barriers; Scattering in One Dimension, II: Square Wells; Scattering in One Dimension, III: Edge Effects; Scattering in One Dimension, IV: Momentum Space.

A series of 9 film loops in b/w. Produced by the Education Development Center (W. Bowers, A. Goldberg, J. C. Hamilton, H. M. Schey, and J. L. Schwartz). Distributed by EDC, Holt/Ealing, Kalmia, and de Gruyter (in **Experiments in Physics Series**). See ordering information for each distributor above.

Computer-animated films that visualize the time development of wave packets in various one-dimensional environments. All of the phenomena are described by the time-dependent Schrödinger equation. (Review: American Journal of Physics, Vol. 46, No. 3, March 1978, p. 315.)

Light--The Wave That Wasn't--Quantum Theory.

Filmstrips from Audio Visual Narrative Arts, Inc., Box 9, Pleasantville, NY 10570. Price: 4 filmstrips and 4 lp records, $107.

Discusses evidence for both the wave and particle models. (Review: Physics Teacher, May 1979, p. 326.)

The Concept That Shook Physics--Quantum Theory.

Filmstrips from Audio Visual Narrative Arts, Inc., Box 9, Pleasantville, NY 10570. Price: 2 filmstrips and 2 lp records, $53.50.

Provides useful historical background in reviewing the accomplishments of classical physics and a simplified explanation of black-body radiation. (Review: Physics Teacher, November 1979, p. 548.)

CHAPTER 25 SOLID-STATE PHYSICS

Fluid Helium: Superconductivity, trapped flux, Meissner-Ochsenfeld effect.
Superfluid Helium: Fountain effect.
Bragg-Nye Crystal Model.

From the **Experiments in Physics: Thermodynamics, Optics, and Electricity Series.** See ordering information above. (Review: American Journal of Physics, Vol. 45, No. 10, October 1977, pp. 1014-1015.)

FILMS

FILM RESOURCE BOOK

A Cinescope of Physics, John Dowling, ed.
Includes hundreds of annotated film entries, analyzed according to physics quality, social relevance, photography, and other parameters. Also includes an explanatory introduction, a listing of the major film companies and distributors, and a bibliography of writings of audio-visual aids in the physics classroom.

FILMS

(listed in order of the chapters in **Invitation to Physics**)

Many of the films listed in this section have been reviewed in the American Journal of Physics. The film editor is John Dowling, Jr., Department of Physics, Mansfield State College, Mansfield, PA 16933. He is always looking for new reviewers.

CHAPTER 1 THE PHYSICS OF THE UNIVERSE

Powers of 10--1978.
Produced by Charles Eames. Distributed by Pyramid Films, Box 1048, Santa Monica, CA 90406. 16 mm, color, animated, 9 min., Revised 1978. Price: $200 purchase, $25 rental.
A cosmic zoom of our universe in powers of 10, up to clusters of galaxies and then down to an atomic nucleus. The revised film is in color and is narrated by Philip Morrison. It features improved coverage of the microscopic world. Some viewers may prefer the original 1968 version. (Review: American Journal of Physics, Vol. 47, No. 3, March 1979, p. 297.)

The House of Science.
Produced by Charles and Ray Eames. Distributed by Encyclopaedia Britannica Education Corp., 425 N. Michigan Ave., Chicago, IL 60611. 16 mm, color, 15 min., 1968. Price: $185 purchase, $18.50 rental.

Presents impressions of science. (Review: American Journal of Physics, Vol. 46, No. 3, March 1978, pp. 316-317.)

World Within World.
Produced by BBC-TV and Time-Life Films. Distributed by Time-Life Multimedia, 100 Eisenhower Dr., Paramus, NJ 07652. 16 mm and 3/4-in. videocassette, color, 52 min. Price: $660 (16 mm) or $465 (video) purchase, $100 rental.

Traces the history of people and ideas that have made 20th-century physics "the greatest achievement of the imagination."

CHAPTERS 2-5 MECHANICS

The following films are distributed by Modern Film Rentals, 2323 New Hyde Park Rd., New Hyde Park, NY 11040.

Collisions of Hard Spheres. Prepared by James Strickland, PSSC. b/w, 19 min. Purchase: $151.

Conservation of Energy. Prepared by Arthur La Croix, New England Electric and Jerrold R. Zacharias, MIT. b/w, 27 min. Purchase: $189.

Free Fall and Projectile Motion (Falling Bodies). Prepared by Nathaniel Frank, MIT. b/w, 27 min. Purchase: $189.

Forces. Prepared by Jerrold R. Zacharias, MIT. b/w, 32 min. Purchase: $151.

Inertia. Prepared by E. M. Purcell, Harvard. b/w, 26 min. Purchase: $189.

Inertial Mass. Prepared by E. M. Purcell, Harvard. b/w, 19 min. Purchase: $151.

A Million to One. b/w, 5 min. Purchase: $189.

Moving with the Center of Mass. Prepared by Hermon Branson, Howard University. b/w, 26 min. Purchase: $189.

Straight Line Kinematics. Prepared by E. M. Hafner, Rochester. b/w, 34 min. Purchase: $189.

CHAPTER 6 ROTATION AND ANGULAR MOMENTUM

Frames of Reference.

Prepared by Patterson Hume and Donald Ivey, University of Toronto. Distributed by Modern Film Rentals, 2323 New Hyde Park Rd., New Hyde Park, NY 11040. b/w, 28 min. Purchase: $189.

Conservation Laws in Zero-g.

Available from NASA. 18 min., 1974. See ordering information at the end of this section.

Dr. Owen K. Garriot, scientist-astronaut, discusses from Skylab and from the classroom back on earth demonstrations of angular momentum, using as some of the examples astronauts spinning and rotating in space, an ice skater spinning, a cat dropped with zero angular momentum landing on its feet, and a model of the Explorer I satellite.

Gyroscopes in Space.

Available from NASA. 15 min., 1974. See ordering information at the end of this section.

Illustrates the basic principles of gyroscopes by utilizing kinescope and other film footage from Skylab. Dr. Owen K. Garriot, a scientist-astronaut, discusses the experiment.

CHAPTER 7 GRAVITY

Newton: The Mind That Found the Future.

Produced and distributed by the Learning Corporation of America, 1350 Avenue of the Americas, New York, NY 10019. 16 mm, color, 21 min. Price: $380 purchase, $30 rental. Released 1971.

Biography of Newton. (Review: American Journal of Physics, Vol. 45, No. 2, February 1977, pp. 231-232.)

Copernicus.

Produced by Charles and Ray Eames. Distributed by Pyramid Films, P.O. Box 1048, Santa Monica, CA 90406. 16 mm, color, 10 min. Price: $145 purchase, $15 rental. Released 1973.

A look at surroundings, artifacts, and documents of Copernicus and how he came to advance the heliocentric theory.

The Starry Messenger.

Produced by BBC-TV and Time-Life Films. Distributed by Time-Life Multimedia, 100 Eisenhower Dr., Paramus, NJ 07652. 16 mm or 3/4-in. videocassette, color, 52 min. Price: $660 (16 mm) or $465 (video) purchase, $100 rental (either format). Released 1974.

No. 6 in **The Ascent of Man** series narrated by Jacob Bronowski. About the development and acceptance of the Copernican theory, including a masterful part about the trial of Galileo. (Review: American Journal of Physics, Vol. 45, No. 3, March 1977, p. 320.)

Elliptical Orbits.

Prepared by Albert V. Baez, PSSC. Distributed by Modern Film Rentals, 2323 New Hyde Park Rd., New Hyde Park, NY 11040. b/w, 19 min. $114.

The Motions of Attracting Bodies.
Planetary Motion and Kepler's Laws.

Distributed by Houghton-Mifflin Co., One Beacon St., Boston, MA 02107. Approximately 8 min. each. Price: approximately $125 each, purchase; approximately $20 each, rental.

The images are generated by a computer to demonstrate these single concepts; electronic music and narration are provided. A pamphlet entitled "Film Notes" by Michael Zeilik II is included.

CHAPTER 8 SPECIAL RELATIVITY
CHAPTER 9 GENERAL RELATIVITY

Relativistic Time Dilation.

Produced by Paul Hewitt. Distributed by Steve Smith Associates, 2341 Grant St., Berkeley, CA 94703. 16 mm, color, animated, 12 min. Price: $170 purchase. Released 1976.

The twin paradox of special relativity. (Review: American Journal of Physics, Vol. 45, No. 10, October 1977, p. 1016.)

Time Dilation--An Experiment with mu-Mesons.
Produced by David H. Frisch, MIT, and James H. Smith, University of Illinois. Distributed by Modern Film Rentals, 2323 New Hyde Park Rd., New Hyde Park, NY 11040. b/w, 36 min. Purchase: $189.

Treats spacetime effects of special relativity, showing an actual experiment. (Review: American Journal of Physics, Vol. 45, No. 10, October 1977, p. 1016.)

An Inertial Box.
Produced by Richard Keiper and Daniel Mason under the direction of Donna Berry. Distributed by AAPT Film Repository, Graduate Physics Building, SUNY, Stony Brook, NY 11794. Super 8 mm Technicolor Cartridge, silent, color, 3 min., 1976. Price: $25 purchase.

Shows the meaning of the term "weightlessness" as applied to a body in free fall. (Review: American Journal of Physics, Vol. 48, No. 7, July 1980, pp. 590-591.)

Gravity Waves: Search and Discovery.
Produced and distributed by American Educational Films, 132 Lasky Dr., Beverly Hills, CA 90212. 16 mm, color, animated, sound, 15 min. Price: $255 purchase, $30 rental. Released 1976.

What gravity waves are and how they can be detected. (Review: American Journal of Physics, Vol. 45, No. 1, January 1977, p. 117.)

The Majestic Clockwork.
Produced by BBC-TV and Time-Life Films. Distributed by Time-Life Multimedia, 100 Eisenhower Dr., Paramus, NJ 07652. 16 mm or 3/4-in. videocasette, color, 52 min. Price: $660 (16 mm) or $465 (video) purchase, $100 rental (either format).

No. 7 in the **Ascent of Man** series narrated by Jacob Bronowski. This film covers the contributions of Newton and Einstein and the revolution that ensued when Einstein's theory of relativity upset Newton's elegant description of the universe.

Einstein.

Produced by Patrick Griffin for NOVA. Distributed by NOVA, WGBH Educational Foundation, 125 Western Ave., Boston, MA 02134. 16 mm or 3/4-in. videocassette, color, 60 min., 1979. Price: $450 (16 mm) or $325 (video) rental. For purchase inquire.

A biography of Einstein. (Review: American Journal of Physics, Vol. 48, No. 7, July 1980, pp. 588-589.)

A Portrait of Albert Einstein.

Produced and distributed by Francis S. Lestingi, Department of Interdisciplinary Science, SUNY College at Buffalo, Buffalo, NY 14222. 3/4-in. videocassette, b/w, 25 min., 1977. Price: send blank 30-min. videocassette plus $15 for duplicating fee.

A biographical sketch of the life of Einstein. (Review: American Journal of Physics, Vol. 46, No. 7, July 1978, p. 780.)

Einstein's Universe.

Produced by Martin Freeth (BBC) and Jo Gladstone (WGBH). Distributed by Corinth Films, 410 East 62nd St., New York, NY 10021. 16 mm or 3/4-in. videocassette, color, 120 min., 1979. Price: $1000 (16 mm) or $600 (video) for 3-year lease, $128 (16 mm) or $100 (video) rental.

Presents Einstein's theories and their impact on our view of the cosmos. Several physicists in the field were assembled at the McDonald Observatory in Texas to make this TV movie, which is narrated by Peter Ustinov. (Review: American Journal of Physics, Vol. 48, No. 7, July 1980, pp. 588-589.)

CHAPTER 10 FLUIDS

The Unusual Properties of Liquid Helium.

Produced and distributed by Isadore Rudnick, Physics Dept., University of California, Los Angeles, CA 90024. 16 mm, color, 17 min., 1977. Price: $300 purchase, $34 rental from Film Services, 860 Second St., San Francisco, CA 94107.

Focuses on the properties of the superfluid

component with many spectacular demonstrations. (Review: American Journal of Physics, Vol. 46, No. 7, July 1978, pp. 780-781.)

Superfluid Helium.

Produced and distributed by J. F. Allen, Dept. of Physics, University of St. Andrews, North Haugh, St. Andrews, Fife KY 16955 Scotland. 16 mm, color, 16 min. Price: $240 purchase. Released 1974.

Demonstrates many of the peculiar characteristics of superfluid helium. (Review: American Journal of Physics, Vol. 46, No. 11, November 1978, p. 1207.)

Fluids in Weightlessness.

Available from NASA. 14 min., 1974. For ordering information, see end of section on Films.

Filmed on Skylab.

CHAPTER 11 TEMPERATURE AND HEAT
CHAPTER 12 HOW HEAT GETS AROUND

Kinetic Theory by Computer Animation.

11 color films. Produced by John T. Fitch. Distributed by Kalmia, Department P5, Concord, MA 01742. 4 min. each. Price: $34.95 for 16 mm. (Also available in Super 8 mm format.)

A series of films developed for the Self-Study Subject Development Center for Advanced Engineering Study at MIT by John T. Fitch and Sherrill F. Martin. (Review: Physics Teacher, March 1980, p. 222.)

CHAPTER 13 WAVES
CHAPTER 14 STANDING WAVES

The following films are distributed by Modern Film Rentals, 2323 New Hyde Park Rd., New Hyde Park, NY 11040.

Simple Waves. Prepared by John Shive, Bell Telephone Laboratories. b/w, 27 min. Purchase: $189.

Interference and Diffraction. Prepared by James Strickland, EDC. b/w, 19 min. Purchase: $113.

Doppler Effect and Shock Waves. Prepared by James Strickland, EDC. b/w, 8 min. Purchase: $63.

The Doppler Effect in Sound and Light.
Distributed by Houghton-Mifflin Co., One Beacon St., Boston, MA 02107. Approximately 8 min. each. Price: approximately $125, purchase; approximately $20, rental.

The images are generated by a computer to demonstrate these single concepts; electronic music and narration are provided. A pamphlet entitled "Film Notes" by Michael Zeilik II is included.

Simple Harmonic Motion.
Produced and distributed by the Graduate Program in Acoustics and the Applied Research Laboratory, Pennsylvania State University, P.O. Box 30, State College, PA 16801. 16 mm, color, 16.5 min., 1978. Price: $200 purchase.

Illustrates and analyzes simple harmonic motion by means of computer animation. (Review: American Journal of Physics, Vol. 47, No. 8, August 1979, p. 754.)

A Look at Sound.
Produced and distributed by Time-Life Films, 100 Eisenhower Dr., Paramus, NJ 07652. 16 mm or 3/4-in. videocassette, color, 30 min. Price: $400 (16 mm) or $200 (video) purchase, $40 rental. Released 1977.

Phenomena of sound and vibration and some everyday applications. (Review: American Journal of Physics, Vol. 46, No. 11, November 1978, p. 1207.)

See: **A Whisper from Space** under Chapter 29; it includes a recreation of an early Doppler effect experiment with sound.

CHAPTER 15 LIGHT--THE WAVE SIDE

Polarization of Single Photons.
Produced by S. Berko. Distributed by Educational Development Center, 39 Chapel St., Newton, MA 02160. 16 mm, color, 15 min. Price: $165 purchase, $15 rental. Released 1972.

Investigates the quantum behavior of polarized

photons. (Review: American Journal of Physics, Vol. 46, No. 3, March 1978, p. 317.)

Interference in Photon Polarization.
Produced by Edwin Taylor. Distributed by Educational Development Center. 16 mm (sale and rental) or Super 8 mm (rental only), silent, color, 4 min. Price: $45 purchase, $10 rental. Released 1973.

Shows polarization and recombinations of doubly refracted light beam using two calcite crystals. (Review: American Journal of Physics, Vol. 46, No. 3, March 1978, p. 317.)

CHAPTER 16 HOLOGRAPHY

The Process of Holography.
Produced by Edwin Taylor. Distributed by Educational Development Center. 16 or Super 8 mm, silent, color, 10 min. Price: $150 purchase.

Illustrates the physical basis for recording a simple hologram and reconstructing the image. (Review: American Journal of Physics, Vol. 46, No. 3, March 1978, p. 317.)

CHAPTER 17 OPTICS
CHAPTER 18 OPTICAL INSTRUMENTS
COLOR SECTION ON COLOR

The following films are distributed by Modern Film Rentals, 2323 New Hyde Park Rd., New Hyde Park, NY 11040.

Introduction to Optics. Prepared by E. P. Little, PSSC. color, 23 min. Purchase: $218.

Reflection and Refraction. Prepared by James Strickland, EDC. b/w, 17 min. Purchase: $113.

Color by Subtraction.
Produced and distributed by Science Software Systems, Inc., 11899 West Pico Blvd., West Los Angeles, CA 90064. 16 mm, color, 24 min., 1978. Price: $350 purchase, $35 rental.

Reviews concepts of color by addition and visualizes the principles of color photography and color printing.

Colors and How We See Them.

Produced and distributed by Science Software Systems, Inc., 11899 West Pico Blvd., West Los Angeles, CA 90064. 16 mm, color, 20 min., 1978. Price: $300 purchase, $30 rental.

Treats the "Young-Helmholtz" three-component theory of color vision.

Color from Light.

Clark Dugger. Churchill Films, 662 North Robertson Blvd., Los Angeles, CA 90069. 16 mm, 11 min., color. Price: $155 purchase, $15 rental.

Is accompanied by a study guide suggesting discussion and activities. (Review: Physics Teacher, January 1979, p. 54.)

Light and Its Relation to Color.

Produced and distributed by Science Software Systems, Inc., 11899 West Pico Blvd., West Los Angeles, CA 90064. 16 mm, color, 20 min., 1978. Price: $300 purchase, $30 rental.

Treats many aspects of the subject.

The Simple Lens: An Introduction.

Produced by Jim Veilleux. At release (1976) distributed by BFA Educational Media, 2211 Michigan Ave., Santa Monica, CA 90404. Inquire of Holt. 16 mm, color, 12 min. Price: $175 purchase, $15 rental.

Explains how a lens refracts light and forms an image. (Review: American Journal of Physics, Vol. 45, No. 5, May 1977, p. 504.)

The Revealing Eye.

Produced by the Shell Oil Co. Distributed by the Shell Film Library, 1433 Sadlier Circle West Dr., Indianapolis, IN 46239. 16 mm, some color, 19 min. Free loan. Released 1973.

Explores the use of the camera as a time machine. (Review: American Journal of Physics, Vol. 45, No. 5, May 1977, p. 504.)

CHAPTERS 19–22 ELECTRICITY AND MAGNETISM

The following films are distibuted by Modern Film Rentals, 2323 New Hyde Park Rd., New Hyde Park, NY 11040.

Coulomb's Law. Prepared by Eric Rogers, Princeton. b/w, 30 min. Purchase: $189.

Electric Fields. Prepared by Francis Bitter, MIT, and John Waymouth, Sylvania Electrical Products. b/w, 25 min. Purchase: $151.

Electric Lines of Force. Prepared by Alexander Joseph, Bronx Community College. b/w, 7 min. Purchase: $52.

A Magnet Laboratory. Prepared by Francis Bitter, MIT, and John Waymouth, Sylvania Electrical Products. b/w, 21 min. Purchase: $151.

Millikan Experiment. Prepared by Francis L. Friedman, MIT, and Alfred Redfield, IBM. b/w, 30 min. Purchase: $189.

Magnetism in Space.
Available from NASA. 14 min., 1975. See ordering information at end of Film section.
Filmed on Skylab.

CHAPTER 23 QUANTUM THEORY AND ATOMIC STRUCTURE

The following films are distributed by Modern Film Rentals, 2323 New Hyde Park Rd., New Hyde Park, NY 11040.

Franck-Hertz Experiment. Prepared by Byron Youtz, Reed College. b/w, 30 min. Purchase: $189.

Interference of Photons. Prepared by John King, MIT. b/w, 13 min. Purchase: $114.

Matter Waves. Produced by Allan Holden and Lester Germer, Bell Telephone Laboratories. b/w, 28 min. Purchase: $189.

Photons. Prepared by John King, MIT. b/w, 19 min. Purchase: $113.

Pressure of Light. Prepared by Jerrold R. Zacharias, MIT. b/w, 23 min. Purchase: $151.

The Stern-Gerlach Experiment. Prepared by Jerrold R. Zacharias, MIT. b/w, 27 min. Purchase: $151.

Knowledge or Certainty.
Produced by BBC-TV and Time-Life Films. Distributed by Time-Life Multimedia, 100 Eisenhower Dr., Paramus, NJ 07652. 16 mm or 3/4-in. videocassette, color, 52 min. Price: $660 (16 mm) or $465 (video) purchase, $100 rental (either format).

No. 11 in the **Ascent of Man** series narrated by Jacob Bronowski. The moral dilemma for mankind as posed in the language of science. (Review: American Journal of Physics, Vol. 45, No. 3, March 1977, pp. 320–321.)

Quantum Physics Series.
16 mm, b/w, silent, computer-animated, 3–4 min. each. Available from EDC and and other distributors. For further ordering information, see listings above. Price: $20 purchase each (EDC), $20 series rental (EDC).

Computer-animated films which visualize the time development of wave packets in various one-dimensional environments. All of the phenomena are described by the time-dependent Schrodinger equation. (Review: American Journal of Physics, Vol. 46, No. 3, March 1978, p. 315.)

CHAPTER 24 LASERS

Lasers: The Light of the 21st Century.
Produced for WGBH-TV for NOVA. Distributed by Time-Life Multimedia, 100 Eisenhower Dr., Paramus, NJ 07652. 16 mm or 3/4-in. videocassette, color, 52 min., 1978. Price: $650 (16 mm) or $325 (video) purchase, $65 rental (either format).

Discusses the physics of lasers, recent medical, industrial, and military applications of lasers, and the controversy of who discovered the laser. (Rev. American Journal of Physics, Vol. 47, No. 3, March 1979, p. 298.)

The Laser.
Produced by Bell Laboratories. 16 mm. 10 min. Can be purchased from M.G.S., 619 W. 45th St., New York, NY 10019. Free loan: write Bell Laboratories, 600 Mountain Ave., Murray Hill, NJ 07974. Their brochures

say that the films can also be borrowed from "your local Bell System telephone company's film library. Consult your local telephone directory for the location and telephone number of your telephone company's headquarters."

Charles H. Townes and Arthur Schawlow discuss events leading to discovery of the laser and developments in the field since that time.

CHAPTER 25 SOLID-STATE PHYSICS

The Electrical Digital Computer.
Magnetic Bubbles.
Charge-Coupled Devices (CCDs).

Produced by Bell Laboratories. 16 mm. 8-10 min. Can be purchased from M.G.S., 619 W. 45th St., New York, NY 10019. Free loan: write Bell Laboratories, 600 Mountain Ave., Murray Hill, NJ 07974. Their brochures say that the films can also be borrowed from "your local Bell System telephone company's film library. Consult your local telephone directory for the location and telephone number of your telephone company's headquarters."

Bubble Model of a Crystal Series.

Distributed by Kalmia. 2 films, b/w. $28.95 each.

Bubble Model of a Crystal: Structure and Boundaries; Bubble Model of a Crystal: Deformation and Dislocations.

CHAPTER 26 INSIDE THE NUCLEUS

The following films are distributed by Modern Film Rentals, 2323 New Hyde Park Rd., New Hyde Park, NY 11040.

Positron-Electron Annihilation. Prepared by Stephen Berko, Brandeis University. b/w, 26 min. Purchase: $189.

Rutherford Atom. Prepared by Robert I. Hulsizer, University of Illinois. b/w, 40 min. Purchase: $189.

Building An Atomic Accelerator.

Distributed by Energy Research and Development Administration, Film Library-TIC, P.O. Box 62, Oak

Ridge, TN 37830. 16 mm, color, 32 min. Free loan. Released 1973.

Physical principles and building of the heavy-ion accelerator. (Review: American Journal of Physics, Vol. 45, No. 1, January 1977, pp. 116–117.)

Search for Antiworlds.

Produced by Berkeley Film Associates. Distributed by Pyramid Films, Box 1048, Santa Monica, CA 90406. 16 mm, color, 25 min. Price: $350 purchase, $30 rental. Released 1976.

About antimatter. (Review: American Journal of Physics, Vol. 45, No. 3, March 1977, p. 319.)

The Ascent of Man.
No. 4: The Hidden Structure.
No. 10: World Within World.

Produced by BBC-TV and Time-Life Films. Distributed by Time-Life Multimedia, 100 Eisenhower Dr., Paramus, NJ 07652. 16 mm or 3/4-in. videocassette, color, 52 min. Price: $660 (16 mm) or $465 (video) purchase, $100 rental (either format).

Two films in the series narrated by Jacob Bronowski. The first film traces our knowledge of the elements, and the second looks at the world within the atom. (Review: American Journal of Physics, Vol. 46, No. 5, May 1978, pp. 593–594.)

Marie Curie--A Love Story.

Produced by Chatsworth Films. Distributed by Centron Films, 1621 West 9th St., Box 687, Lawrence, KS 66044. 16 mm, color, 32 min., 1978. Price: $535 purchase.

Biography of Marie Curie.

CHAPTER 27 ENERGY FOR TOMORROW

Which Energy?

Produced and distributed by Stuart Finley, Inc., 3428 Mansfield Road, Falls Church, VA 22041. 16 mm, color, sound, 23 min. Free purchase preview. Price: $350 purchase, $35 rental. Released 1976.

Best comprehensive introduction. Which sources of energy should be pursued and what are the related timing and environmental considerations to be faced. Study guide provided with film. (Review: American Journal of Physics, Vol. 45, No. 8, August 1977, p. 784.)

Energy: The Problems and the Future.
Energy: The Fuels and Man.

Produced and distributed by National Geographic Educational Services, 17th and M Sts., NW, Washington, DC 20036. 16 mm or 3/4-in. videocassette, color, 23 min., 1978. Price: $310 purchase each (either format).

The first film examines the potential of some renewable resources. The second film addresses the questions: What is energy? How is it obtained?

The Invisible Flame.

Coproduced by BBC-TV and WGBH-TV for NOVA. For distribution information, contact Distribution Office, WGBH Educational Foundation, 125 Western Ave., Boston, MA 02134. 3/4-in. videocassette, color, 60 min., 1979.

A look at the potential of hydrogen as a zero-pollution fuel. (Review: American Journal of Physics, Vol. 48, No. 7, July 1980, p. 587.)

Geothermal Power.

Produced and distributed by Extension Media Center, University of California, Berkeley, CA 94720. 16 mm, color, 30 min. Price: $360 purchase, $23 rental. Released 1970.

Examines history and operation of geyser power stations and discusses the future of geothermal power.

Geothermal Energy from Hot Dry Rock.

Produced and distributed by Los Alamos Scientific Lab, Attn: Group Leader, ISD-9, P.O. Box 1663, Mail Stop 415, Los Alamos, NM 87545. 16 mm, color, 13.5 min., 1978. Free loan.

Explains LASL's method for extraction of geothermal energy.

FILMSTRIPS AND FILMS

The Bell Solar Cell.

Produced by Bell Laboratories. 16 mm. 10 min. Can be purchased from M.G.S., 619 W. 45th St., New York, NY 10019. Free loan: write Bell Laboratories, 600 Mountain Ave., Murray Hill, NJ 07974. Their brochures say that the films can also be borrowed from "your local Bell System telephone company's film library. Consult your local telephone directory for the location and telephone number of your telephone company's headquarters."

The Solar Promise.

Produced by Henry Mayer, M.D. Distributed by Bullfrog Films, Oley, PA 19547. 16 mm or 3/4-in. videocassette, color, 28 min., 1979. Price: $390 purchase, $38.50 rental.

Examines the use of solar for heating purposes and covers some of the physics involved. (Review: American Journal of Physics, Vol. 48, No. 7, July 1980, pp. 587-588.)

A Building in the Sun.

Produced by Griffen Productions. Distributed by Bullfrog Films, Oley PA 19547. 16 mm or 3/4-in. videocassette, color, 20 min., 1979. Price: $920 purchase, $31 rental.

Follows the construction of the Plant Science Building of the Cary Arboretum. (Review: American Journal of Physics, Vol. 48, No. 7, July 1980, pp. 587-588.)

Solar Energy--The Great Adventure.

Produced by DOE. Distributed by National Audiovisual Center, Washington, DC 20409. 16 mm, color, 28 min., 1978. Price: $162.50 purchase.

Explores many ways of utilizing solar energy.

A Play Half Written.

Produced by Bill Stokes Associates for the Atomic Industrial Forum. Distributed by Atomic Industrial Forum, 7101 Wisconsin Ave., Washington, DC 20014. 16 mm, color, 26 min., 1979. Price: $125 purchase.

Presents the industry's case for nuclear power.

End of the Rainbow.

Produced by WGBH-TV for NOVA. For distribution, contact Distribution Office, WGBH Educational Foundation, 125 Western Ave., Boston, MA 02134. 3/4-in. videocassette, color, 60 min., 1979.

Examines the promise and problems of fusion as a future energy source. (Review: American Journal of Physics, Vol. 48, No. 7, July 1980, p. 587.)

Fusion: The Energy Promise.

Coproduced by NOVA and BBC-TV. Distributed by Time-Life Multimedia, 100 Eisenhower Dr., Paramus, NJ 07652. 16 mm, color, 52 min. Price: $550 purchase, $60 rental. Released 1974.

Traces the history of the international race to achieve fusion with an explanation of the scientific principles involved. (Review: American Journal of Physics, Vol. 45, No. 7, July 1977, pp. 691-692.)

Laser Fusion.

Produced and distributed by Los Alamos Scientific Lab, Attn: Group Leader, ISD-9, P.O. Box 1663, Mail Stop 415, Los Alamos, NM 87545. 16 mm, color, 15 min., 1977. Free loan.

Explains the techniques and principles of the laser fusion program under development at LASL.

Target for Antares.

Produced and distributed by Los Alamos Scientific Lab, Attn: Group Leader, ISD-9, P.O. Box 1663, Mail Stop 415, Los Alamos, NM 87545. 16 mm, color, 16 min., 1978. Free loan.

Uses microcinematography and animation to describe the fabrication of targets to be used to generate energy through laser fusion techniques.

CHAPTER 28 QUARKS: AN ASSORTMENT OF COLORS AND FLAVORS

Discovery of the "J" Particle.

Produced and distributed by Center for Engineering Studies, MIT, Dept. 19, Rm. 9-234, 77 Massachusetts Ave., Cambridge, MA 02139. 3/4-in. videocassette, color, 80 min., 1978. Price: $510 purchase.

An address by Nobel Laureate S. C. C. Ting in which he describes the discovery of the new "J" particle. Includes a 22-min. introduction by V. Weisskopf.

The Key to the Universe.
Coproduced by BBC and WTTW (Chicago). Distributed by Films Incorporated, 1144 Wilmette Ave., Wilmette, IL 60091. 3/4-in. videocassette, color, 120 min. Price: $925 purchase, $75 rental. Released 1977.

An excellent film by Nigel Calder about elementary particles, quarks, and black holes. A book form is available. (Review: American Journal of Physics, Vol. 45, No. 12, December 1977, pp. 1238-1239.)

CHAPTER 29 ASTROPHYSICS

The Universe--Man's Changing Perceptions.
Produced and distributed by Howard Preston Films, 920 Centinela Ave., Santa Monica, CA 90403. 16 mm, color, 29 min. Price: $400 purchase, $40 rental. Released 1976.

An excellent and scientifically accurate film. (The review--American Journal of Physics, Vol. 45, No. 11, November 1977, pp. 1136-1137--is very negative but we disagree. Perhaps the difference is that the reviewer considered only the film as a stand-alone entity, while if it is shown in the context of a course, many of his objections do not apply.)

Two Hundred Years of American Astronomy.
Produced by Richard Berendzen. Distributed by Director of Programming, The American University, Washington, DC 20016. 3/4-in. videocassette, color, 30 min. Price: $35 purchase, $5 rental. Released 1976.

History of American astronomy and projections into future.

Universe.
Available from NASA. 28 min., 1976. Rental: Free. See ordering information at the end of this section.

An overview of the range of astronomical objects. The film has won prizes, but there are too many errors for my tastes. The rental price is right, however.

"Cosmos" by Carl Sagan: A Personal Voyage.

Off-air film license available from The Television Licensing Center, 1144 Wilmette Avenue, Wilmette, IL 60091.

Birth and Death of a Star.

Distributed by Time-Life Multimedia, 100 Eisenhower Dr., Paramus, NJ 07652. 35 min., 1974. Price: $40 rental from Time-Life or $10 rental from the Astronomical Society of the Pacific, 1244 Noriega St., San Francisco, CA 94122..

An excellent film about the origin, evolution, and death processes of stars. Stellar structure and nuclear reactions are also discussed.

The Sunspot Mystery.

Produced by WGBH-TV as part of NOVA series. Distributed by Distribution Office, WGBH-TV, 125 Western Ave., Boston, MA 02134. 3/4-in. videocassette, color, 59 min. Released 1977.

Examines possible connection between sunspots and the earth's climate. (Review: American Journal of Physics, Vol. 45, No. 8, August 1977, pp. 784-785.)

Solar Eclipse '73.

23 min. Distributed by Modern Film Rentals, 2323 New Hyde Park Rd., New Hyde Park, NY 11040. Can also be ordered directly from National Geographic Society, Washington, DC 20036. Price: approximately $310 purchase; $20 for 3 days, $40 for 8 days rental.

Film comes with associated printed materials and directions for an in-class experiment for construction of a small spectroscope and information about getting spectroscopes and Fraunhofer lines. One of each of the following comes with the film:

"Sun and Light," a booklet,

A device to illustrate spectra of the sun and laboratory devices,

Teacher's Guide,

A piece of diffraction grating.

Additional copies of the associated materials are

available for $1.50. Rental includes one nonreturnable kit.

The Origin of the Elements.

Produced by Dennis Sanders. Distributed by CRM/McGraw-Hill Films, 110 15th St., Del Mar, CA 92014. 16 mm, color, 18 min. Price: $295 purchase, $30 rental. Released 1973.

Explores the nuclear and stellar processes involved in creating the elements; portrays life cycle of stars.

Three Degrees.

Produced by Bell Laboratories. 25 min. Free loan from Modern Talking Picture Service, 500 Park St. N., St. Petersburg, FL 33709. For purchasing information, contact M.G.S., 619 W. 54th St., New York, NY 10019.

The story of Penzias and Wilson's discovery.

A Whisper from Space.

Produced by BBC/WGBH for NOVA. Distributed by Time-Life Multimedia, Room 32-48, Time & Life Bldg., New York, NY 10020. 16 mm or 3/4-in. videocassette, color, 57 min., 1978. Price: $650 (16 mm) or $325 (video) purchase, $65 rental (either format).

Examines the important questions in cosmology--is the universe expanding? was there a big bang? etc. Narrated by Philip Morrison of MIT.

The following are the Apollo Mission films, available from NASA. The later ones emphasize the scientific aspects more than the earlier ones do.

Eagle Has Landed: The Flight of Apollo 11. 29 min., 1969.

Apollo 12: Pinpoint for Science. 28 min., 1969.

Apollo 13: "Houston . . . We've Got a Problem." 28 min., 1970. The one that didn't make it to the moon. The story of the crisis.

Apollo 14: Mission to Fra Mauro. 28 min., 1971.

Apollo 15: In the Mountains of the Moon. 28 min., 1970.

Apollo 16: Nothing So Hidden. 28 min., 1972.

Apollo 17: On the Shoulder of Giants. 28 min., 1973.

The Mission of Apollo–Soyuz. 29 min., 1975.

The following are Skylab and Space Shuttle movies, available from NASA:

Opportunities in Zero Gravity. 28 min., 1976.
Skylab: The First 40 Days. 25 min., 1973.
Skylab: The Second Manned Mission—A Scientific Harvest. 37 min., 1974.
Skylab: Space Station 1. 28 min., 1974.
Spaceship Skylab: Wings of Discovery. 9 min., 1974.

CHAPTER 30 PHYSICS NOW

The Flight of the Gossamer Condor.
Produced by Shedd Productions. Distributed by Churchill Films, 662 N. Robertson Blvd., Los Angeles, CA 90060. 16 mm, color, 27 min., 1978. Price: $370 purchase.

Gives a filmed history of the development and successful flight of the first person–powered aircraft. (Review: American Journal of Physics, Vol. 47, No. 8, August 1979, pp. 753–754.)

A Fundamental Scientist.
Produced by Bell Laboratories. 16 mm, color, 25 min. Free loan from Modern Talking Picture Service, 500 Park St. N., St. Petersburg, FL 33709. For purchasing information, contact M.G.S., 619 W. 54th St., New York, NY 10019.

Features theoretical physicist Philip Anderson, co–recipient of the 1977 Nobel Prize.

* * * * *

FILMSTRIPS AND FILMS

NASA REGIONAL FILM LIBRARIES

All films for rental from NASA are free of charge, except for return postage.

If you live in:	Write to:
Connecticut Maine Massachusetts New Hampshire New York Rhode Island Vermont	National Audiovisual Center Washington, DC 20409 They forward the request to: RHR Filmedia, Inc. 1212 6th Ave. New York, NY 10036
Alaska* Arizona California Hawaii** Idaho Montana Nevada Oregon Utah Washington Wyoming	NASA Ames Research Center Public Affairs Office Moffett Field, CA 94035
Alabama Arkansas Iowa Louisiana Mississippi Missouri Tennessee	NASA George C. Marshall Space Flight Center Public Affairs Office Marshall Space Flight Center, AL 35812
Delaware District of Columbia Maryland New Jersey Pennsylvania	NASA Goddard Space Flight Center Public Affairs Office Code 202 Greenbelt, MD 20771

FILMSTRIPS AND FILMS

Florida Georgia Puerto Rico Virgin Islands	NASA John F. Kennedy Space Flight Center Code PA-EPB Kennedy Space Center, FL 32899
Kentucky North Carolina South Carolina Virginia West Virginia	NASA Langley Research Center Langley Station Public Affairs Office Mail Stop 154 Hampton, VA 23655
Illinois Indiana Michigan Minnesota Ohio Wisconsin	NASA Lewis Research Center Office of Educational Services 21000 Brookpark Rd. Cleveland, OH 44135
Colorado Kansas Nebraska New Mexico North Dakota Oklahoma South Dakota Texas	NASA Lyndon B. Johnson Space Center Photographic Technical Lab. Audiovisual Office Code JL-13 Houston, TX 77058

*Alaska requestors may also obtain selected NASA films from:

Alaska State Library
Pouch G
Juneau, AK 99801

**Requests for NASA films within the Hawaiian Islands can also be referred to:

The Audiovisual Service Section
Hawaii State Department of Education
Kahala Office Center, Room 103
4211 Waialae Ave.
Honolulu, HI 96816

ANSWERS TO QUESTIONS AT ENDS OF CHAPTERS

Chapter 1--The Physics of the Universe

1. Smallest objects, about 0.1 mm = 10^{-4} m.
   Largest objects, about 1 km = 10^{3} m.
   (Largest)/(Smallest) = $(10^{3})/(10^{-4})$
   = 10^{7}.
2. 3×10^{10} cm/s = $(3 \times 10^{10}$ cm/s$) \times (1$ km/10^{5} cm$)$
   = 3×10^{5} km/s.
3. Multiply the two together to get
   $(6.03 \times 10^{23}) \times (1.67 \times 10^{-24}$ g$) = 1.0$ gram.
4. A chair is made out of molecules. A molecule is made out of atoms. Atoms contain protons and neutrons. The protons and neutrons are made up of quarks.

Chapter 2--Particles in Motion

1. Kinematics--The description of the motion of objects.
   Dynamics--The study of the relationship between forces and motion.
2. For a football, the motion from the passer to the receiver is translation. The spin on the ball is rotation. For a diver, the motion from the board to the water is translation. All of the tumbling and twisting is part of rotation.
3. (a) It is traveling fastest during the second leg, when it goes 40 km in half an hour. (b) Average speed = 120 km ÷ 3.5 h = 34.3 km/h. (c) Going faster than average in first, second, and last legs. (d) Going slower than average in third and fourth legs.
4. (a) Velocities are not equal, because the directions of motion are not the same. (b) The speeds of the two cars are equal; they have the same numerical value, independent of direction.
5. The average velocity is zero. The ball ended up where it started, so the total change in position is zero.

6. 40 km/h east is an arrow to the right.
   (a) 40 km/h north is an arrow the same length upward.
   (b) 20 km/h east is an arrow half the length to the right.
   (c) The 20 km/h west component is a half-length arrow to the left; the 20 km/h south component is a half-length component downward; the resultant is a diagonal from upper right to lower left.
7. You cannot do it. To go 2 km at an average speed of 60 km/h, the trip must take 2 min. Those 2 min have already been used up in going the first km at 30 km/h.
8. You would have to know the total distance traveled, and the total time it took for the trip.
9. First find the average speed in each 1-s interval:

Interval	Avg. Speed
1 to 2 s	1 m/s
2 to 3 s	3 m/s
3 to 4 s	5 m/s
4 to 5 s	7 m/s

   In each 1-s interval, the speed increases by 2 m/s. Therefore, the acceleration must be 2 m/s^2.
10. In Fig. 2-21 the car has the greatest acceleration when it first starts out.
11. On the moon there is no air resistance. This means that the ball and feather should fall with the same acceleration. Note that a film loop is available on which the falling feather and falling hammer are much easier to see than they are in the still pictures in the text.
12. Distances after 0, 1, 2, 3, 4, 5, 6 s--0, 0.83, 3.33, 7.50, 13.33, 20.8, 30.0 m. (Acceleration is 1.7 m/s^2.)
13. The horizontal speed does not depend on the vertical speed.
14. The ball lands in front of the car. It has the same horizontal speed that the car had before the brakes were applied.
15. Relative speed is greatest (70 km/h) when the trains are going in opposite directions.
16. Flying against the wind gives the greatest air speed for a given ground speed.

ANSWERS TO QUESTIONS

Chapter 3--Laws of Motion

1. A straight line with a constant speed.
2. No. If only the earth's gravity were acting, it would go in a curved path. There must be another force to cancel the effect of the earth's gravity, leaving no net force. Over a short distance, however, a gradual curve caused by gravity might not be detectable.
3. It is difficult to find an object with no forces acting on it. Frictional forces are always present.
4. Gravitational force acts downward. This slows the upward motion of the ball, and then increases the downward motion of the ball. Also, the force of air resistance, always in a direction opposite to the motion of the ball, always tends to slow the ball down.
5. (a) Since the mass is doubled, the acceleration is halved, to 0.5 m/s^2. (b) 1 N.
6. (a) 50 N. (b) 0.05 m/s^2.
7. The ground pushing up on your feet (or shoes).
8. (a) Forces are gravity acting downward, and the force of the bar on your hands, acting upward. (b) Forces are your hands pulling downward, the weight of the bar acting downward, and the forces of the walls on the bar, acting upward.
9. 750 N (by Newton's third law).
10. The gas leaving the rear of the rocket.
11. The two forces act on different objects. One acts on you and the other acts on the rink. In looking at the force on you, only one comes into play.
12. A force must be exerted to overcome friction.
13. 100 N.
14. (a) 750 N. (b) 750 N.
15. (a) Weight on moon is 1/6 that on earth. (b) Mass on moon is same as that on earth.
16. Less than 1 g. If it were equal, the ball would fall at a 45° angle. If it were greater, the ball would fall in a path that was even closer to the horizontal.
17. The people cannot stand on the floor. They are in free fall, and would be experiencing weightlessness.
18. Take an object out of your pocket and throw it one way. You will go the other way.
20. The forces are acting at all times, the position and velocity of the object at some particular time.

ANSWERS TO QUESTIONS

Chapter 4--Momentum

1. The feather has the greater momentum, since it is moving. The elephant is standing still, so its momentum is zero.
2. (a) Since both receive the same impulse, both have the same momentum. (b) Since both have the same momentum, the one with the lower mass, the flea, will have the greater speed.
3. Bullet: momentum = 3.0 kg•m/s. Car: momentum = 5000 kg•m/s. The momentum of the car is much greater.
4. Stopping distance is proportional to the square of the speed. Speed is 4 times as great, so stopping distance is 16 times as great, or 320 m.
5. The net force on the object is zero.
6. The scale must be exerting an extra force, beyond the weight of the sand, to stop the falling sand. The falling sand must be exerting an equal and opposite force on the scale, increasing the reading slightly.
7. They must have the same momentum. The 50-kg skater must have 1.5 times the speed of the 75-kg skater.
8. The force that the earth and moon exert on each other is an internal force. All other forces, such as the force of the sun on the earth and moon, are external forces. (b) The momentum of the earth-moon system is not constant, since there are external forces acting.
9. (a) They must have the same momentum, so the speed of the 100-kg skater must be half that of the 50-kg skater, or 1 m/s. (b) The center of mass was initially at rest and will remain at rest, since there were no external forces acting.
10. The car would move a small amount in the opposite direction to the ball's motion. When the ball is moving to the left, the car is moving to the right. When the ball is moving to the right, the car is moving to the left. When the ball is standing still, the car is standing still. This is because the total momentum of the car plus the ball must be zero. If the car is initially in motion, these changes would be superimposed on the initial motion.
11. (a) As you move a brick, say, from left to right, the car moves from right to left so that the center of mass of the system including the car plus the bricks stays

in the same place. (b) The final position of the car only depends on where the bricks are, not how fast they got there. (c) You could throw the bricks with the same result.

12. The air molecules exert a force that is perpendicular to the surface of the wedge, so the force is both downward and backward.
13. (a) As more fuel is burned, the mass of the rocket decreases. As a result, you get more acceleration for a given force. (b) As long as the thrust of the rocket is constant, the momentum change of the rocket in any time interval will still be the same.
14. A horseshoe.
15. The only external force acting on the rocket plus fuel and gas system is the force of gravity. The center of mass of this system will "fall" freely under gravity. (Note that there is an additional complication of the gases striking the earth at launch. This results in an initial forward momentum for this system, instead of a zero momentum.)
16. (a) No, assuming the balls are not tied down. (b) Momentum would not be conserved.
17. (a) No. (b) Before the collision the total momentum would be zero and afterward there would be some momentum, so momentum would not be conserved.
18. (a) The mass of the earth must be about 10^{23} times the mass of a person. (b) If a person has a mass of 75 kg, then the mass of the earth is about 7.5×10^{24} kg. (The actual value is 6×10^{24} kg.)
19. The wobbling motion must be caused by the gravitational effects of something, even if we can't see that something directly.

Chapter 5--Energy

1. It gives us an easier way of looking at many situations.
2. It would be the same. The change in potential energy only depends on the initial and final heights and not on the path taken in between.
3. Twice as much.
4. Method 1: Doing work at a rate of (10 N)(1 m/s), or, 10 J/s. After 10 s, you do 100 J of work.

Method 2: Distance covered is 10 m, so work is force x distance, or 100 J.

5. The lower mass ball will have the greater kinetic energy, since it will have the greater speed.
6. Both will have the same speed, so the more massive ball will have the greater kinetic energy.
7. You will do the same amount of work on the suitcase. In each case, you do the amount of work required to increase the potential energy of the suitcase by the amount corresponding to raising it from one floor to the next.
8. Both have the same potential energy. The potential energy only depends on where it is, not how it got there.
9. 200 J. The gain in kinetic energy is equal to the loss of potential energy. If it falls twice as far, it loses twice as much potential energy.
10. 8 m/s. The change in kinetic energy will just depend on the distance fallen, and not on the location of the starting point.
11. The kinetic energy of the earth will be 10^{-23} times that of the ball. (Since momentum, mv, must be conserved, the velocity of the earth is lower than that of the ball by the same factor that its mass is higher. The mass of the earth is 10^{23} greater than that of the ball, so the velocity that the earth gets is 10^{23} times less. K.E. has one factor of mass and two of velocity. $10^{23} \times 10^{-23} \times 10^{-23} = 10^{-23}$.)
12. The faster object has 4 times the kinetic energy since it has twice the speed.
13. The object with twice the mass has twice the kinetic energy.
14. The kinetic energies are the same. Kinetic energy depends on the speed of the object and not on the direction of motion.
15. (a) Since one car is moving with 3 times the speed of the other, it must have 9 times the kinetic energy of the other, so 9 times the amount of work is required to bring it to a stop. (b) Since the braking forces are the same, this means that 9 times the distance is required to stop the faster car.
16. On a large scale, mechanical energy can disappear in friction. However, this goes into heating the objects

in contact, and the heating is just an increase in the kinetic energy in the microscopic motion of the particles that make up the larger objects.

17. Gravity. The string always exerts a force that is perpendicular to the direction of motion, and it can therefore do no work.
18. The push would give the pendulum extra energy, and it would come back to its original position with that extra energy, meaning that it would continue past the starting position and strike the lecturer.
19. More work is required, since the force is greater once the spring is stretched a little bit.
20. Not elastic. Some of the energy went into breaking up the ball.
21. The ball will have its initial speed, plus twice the speed of the bat, so it will go at 80 m/s.
22. No. The method works by transferring some of the kinetic energy of Jupiter to the spacecraft. If there is no kinetic energy, then there is none to transfer and the method doesn't work.

Chapter 6--Rotation

1. The direction of motion is changing, so the velocity, which is a vector, is changing.
2. Doubling the speed will make the acceleration of the ball four times as great, so the tension will have to be four times as great to keep the ball going in circular motion.
3. (a) Same. (b) Force on the truck is twice that on the car because it has twice the mass.
4. Must be able to provide the linear acceleration to get the ball moving quickly, and must have the strength to hold onto the wire with the great tension necessary to produce the circular motion.
5. Want to increase the acceleration by factor of 2. Speed must increase by square root of 2 to 42.4 m/s.
6. The carseat exerts the force on the baby that brings the baby around the turn.
7. You will continue until you hit the right side of the car.
8. There is less friction between the tires and the road on

a rainy day, so the road can't exert as great a force on the car.

9. The storm starts out farther from the earth's axis than New York is, so it is moving around faster from west to east. It therefore gets ahead (east) of New York.
10. Halved.
11. The acceleration increases, so the force must increase.
12. Half as far from the point of support.
13. It gives you a larger lever arm so you can exert a greater torque on the nut.
14. You must exert the force 10 times as far from the point of support as you place the rock. You must push down 10 times as far, or 1 m.
15. As the spider moves out, the moment of inertia of the spider and turntable increases. Since the angular momentum is fixed, the rotation of the turntable will slow down.
16. Near the edge, since this will produce the greater torque.
17. The effect is increased since the change in moment of inertia will be even greater than without the weights.
18. No external torques acting.
19. Revolution is motion around another object, while rotation is the spinning around an axis for a particular object.
20. If the solar system formed out of a rotating cloud, then conservation of angular momentum means that all of the cloud will continue to rotate in the same direction. If the cloud then breaks up into planets, they will continue to move in the same direction.
21. As the star contracts it rotates faster to form the pulsar in the first place. Also, changes in the radius, through starquakes, cause changes in the rotation period.
22. As the pulsars lose energy, the kinetic energy of rotation is decreased. This means that the rotation must slow down.
23. Increases.

ANSWERS TO QUESTIONS

Chapter 7--Gravity

1. It means that the effects of gravity stretch over great distances, and that the law of gravity in one part of the universe is the same as in other parts of the universe.
2. The circular motion of the moon meant that the moon must be accelerating, and there must be a force to produce that acceleration.
3. (a) We now tend to think that simple elliptical motion is simpler than motions with epicycles. (Copernicus, though, got rid of only some of the epicycles.) (b) In general, if two theories explain a set of phenomena equally well, we adopt the simpler theory.
4. The phases of Venus could not be explained in the Ptolemaic theory. Also, the discovery that moons were orbiting Jupiter showed that things could orbit bodies other than the earth.
5. Newton's laws. Kepler's laws just describe how planets move, but Newton's laws describe underlying causes.
6. Fastest.
7. 5.2 astronomical units. Square the period and take the cube root of the result.
8. 29.3 years. Cube the distance and take the square root.
9. If t is the period,

   $$(t/1.7 \text{ days})^2 \times (15 \text{ radii}/5.9 \text{ radii})^3 = 16.43$$

   $$t = 4.05 \text{ days (earth days)}.$$

   (b) 1 earth day is (4.2/1.7) = 2.47 Jupiter days. The period is then 10.0 Jupiter days.
10. The gravitational forces that two objects exert on each other are equal and opposite.
11. Without it he could not calculate the gravitational force exerted by an extended object, such as the earth.
12. The distances over which these bodies move is small compared to the radius of the earth, so the gravitational force doesn't change much.
13. The earth and the sun are electrically neutral, so there is no net electrical force. This results from there being two types of electrical charge. However, the gravitational forces are only attractive, and just keep adding up as the objects get more massive.
14. To take advantage of the extra kinetic energy provided by the earth's rotation.

ANSWERS TO QUESTIONS

15. No. If you take a strict definition of "synchronous" as meaning that the satellite always stays exactly over one place on the earth, then it must be over the equator. If you relax the definition a bit to mean only that the satellite has a 24-hour orbit, then the satellite can have an orbit inclined to that of the equator but must go both north and south of the equator, and so still can't hover right over New York City. Still, the satellite may remain in view all the time from a given location even if it isn't right overhead. The International Ultraviolet Explorer spacecraft, for example, can always be tracked from NASA's Goddard Space Flight Center in Greenbelt, Maryland, even though the spacecraft actually traces out an ellipse on the ground.
16. No. You are being dragged along by the earth's surface and you are not in free fall as a satellite is.
17. The escape velocity. If satellites went faster than the escape velocity, they would escape.
18. The orbiting moon causes the wobble. It is the center of mass of the earth and moon that goes in the ellipse.
19. The motion of Uranus was not quite as expected, and it was thought that the deviations from the expected path were caused by the gravitational effects of an unseen planet.
20. The gravitational potential energy of the cloud decreases as it gets smaller, so this energy is available to appear as heat.
21. Look for "wobbling" in the motion of the star.
22. On the side of the earth closest to the moon, the moon pulls the water away from the earth. On the other side, the moon pulls the earth away from the water.
23. This is due to the orbital motion of the moon.
24. Nearer, inside the Roche limit.
25. We think that the ring is a moon that could not form under the influence of large tidal forces.
26. The earth and moon would move off in a straight line.

ANSWERS TO QUESTIONS

Chapter 8--Special Relativity

1. They agree when things are moving at speeds much less than that of light.
2. $(5\ m)/(3 \times 10^8\ m/s) = 1.7 \times 10^{-8}\ s$.
3. $1\ ly = 10^6 \times (3 \times 10^8\ m/s) \times (3.1 \times 10^7\ s) = 10^{16}\ m$.
4. $(1\ m)/(3 \times 10^8\ m/s) = 3.3 \times 10^{-9}\ s$.
5. Astronomical observations allow the timing of signals over very large distances, so convenient time intervals are involved.
6. No. The observation of aberration meant that if the ether exists, the earth must be moving through it, and the Michelson-Morley experiment showed that the earth is not moving through the ether.
7. It involves reference frames that are not accelerating.
8. If a unique "universal" time can be assigned to each event, then events that are simultaneous for one observer must be simultaneous for all observers.
9. 0.5 s. The situation is completely symmetric.
10. 0.5 m. The situation is completely symmetric.
11. It is just time as kept by a clock that is at rest in a particular reference frame.
12. 0.9 s.
13. 0.9 times the speed of light.
14. (a) 111 yr. (b) 48 yr.
15. Not even light from the first event could reach the location of the second event before the second event took place, so there is no way for the first event to influence the second event.
16. The kinetic energy is much less than the rest energy, which is greater than that of an atomic bomb.

Chapter 9--General Relativity

1. It is not restricted to inertial frames.
2. Gravitational mass is a measure of the ability of an object to exert and feel gravitational forces. Inertial mass is a measure of the ability of an object to resist being accelerated when a force is applied.
3. If one were a constant, multiplied by the other, then two objects would have the same acceleration. Otherwise they would not.

4. The astronauts and the space station are "falling" with the same acceleration, so the astronauts are weightless relative to the space station.
5. (1) Advance of the perihelion of Mercury. (2) Bending of starlight around the sun. (3) Gravitational redshift on earth. (4) Bending and time delay of signals from spacecraft going behind sun. (3 of the above)
6. (a) One must measure the apparent positions of stars when the sun is near them in the sky. If there were no eclipse of the sun, it would be too bright to see the stars. (b) Distant radio sources can be used, and an eclipse is not necessary.
7. About 1%.
8. There is no direct evidence of our having detected gravity waves.
9. Observations of the binary pulsar suggest that that system is radiating energy in the form of gravitational waves, but this evidence is indirect since we can't detect the gravitational waves.
10. Bending of radio signals from quasars. Advance of perihelion in orbiting objects.
11. Carrying atomic clock in airplane. Gravitational redshift in signal going from basement to roof of building.

Chapter 10--Fluids

1. 3000 kg.
2. 5000 kg/m^3. The density depends on the material, not on how large a sample you take.
3. 2000 kg/m^3.
4. 2×10^5 N.
5. 1×10^5 N/m^2. The pressure does not depend on the size of the table.
6. The temperature and density of the gas.
7. The pressures inside and outside are equal.
8. Same amount of work.
9. Lift in a garage. Brakes in a car. Controls in an airplane.
10. 10^4 N/m^2; the same.
11. 10^4 N; the same.
12. Neither; the buoyant force is the same.

13. The heavier freighter experiences the greater buoyant force.
14. In the salt water it rides higher in the water. Salt water is denser, and therefore more buoyant than fresh water.
15. The density decreases, since the volume is increasing and the mass remains the same.
16. Yes.
17. Remain the same. The amount of ice that is below the water is just equal to the volume of water you get when the ice melts.
18. The tires flatten a little so there is more surface in contact with the road.
19. About 80% of the sea level pressure.
20. You have reduced the pressure inside the cup and the atmospheric pressure outside keeps the cup on your mouth.
21. We would have to use a column of water that is 13.6 times as high as the mercury column.
22. The pilot must know the pressure on the ground at the airport, and usually gets this information by radio.
23. The wind will be stronger in a narrow street. This is because the air flows faster through narrow openings.
24. You have made the hole smaller to make the water come out faster.
25. The pressure will be lower where the air is flowing faster, on the west side, so the ball is pushed to the west.
26. If the ratio of the speeds remains fixed, their difference will increase as they increase.
27. The lift acts as a pressure. The greater the area it has to act over, the greater the force.

Chapter 11--Temperature and Heat

1. The gaps allow the sections of road to expand in warm weather without running into each other.
2. 4 mm. Since the bar is twice as long, it will expand by twice as much.
3. The bulb can hold a large volume of mercury, so when the mercury is heated the volume increases by a significant amount. The only place the extra volume can go is into

the long tube. The narrower the opening in the tube, the more the length of the column will change to bring about the required volume change.

4. 0.51 m. It is half as wide as it is long so the width changes by half the amount that the length changes.
5. It will bend the other way.
6. (a) Twice as much, since the temperature change is twice as great. (b) The same amount. The temperature change is half as large, but the sample is twice as large.
7. (a) Same specific heats, since they are made out of the same material. (b) The larger block has twice the heat capacity.
8. The block with the large specific heat will have the larger heat capacity.
9. Close to the temperature of the ocean. The ocean has so much water that it has a very large heat capacity. It can therefore absorb the heat given off by the horseshoe with very little temperature change.
10. They must be equal.
11. The same. The internal energy depends only on the temperature of the gas, not on how it got to be that way.
12. (a) Internal energy increases because work is being done on the gas, and no heat is flowing in or out. (b) The temperature increases since the internal energy increases.
13. All of it. This is because the energy of the gas must remain constant, since the temperature remains constant.
14. Pressure doubles.
15. Temperature doubles.
16. 400 K and 250 K. This gives us the greatest ratio of high temperature to low temperature. It therefore gives the greatest efficiency.
17. 4 J. There is no net change in energy in a full cycle, so the work done must equal the net amount of heat in (10 J – 6 J).
18. Yes. Heat cannot flow from a cold object to a hot object.
19. Thermal pollution is the heating of the environment by the heat that is not converted into mechanical energy.
20. Yes.
21. Even below the boiling point, some molecules have more than the average energy, and can break free of the

water.

22. The hot skillet makes the water evaporate very quickly. Gas (water vapor) is being formed in a layer between the drop and the skillet, and the drop floats on the gas that is rushing off.
23. The hot runners melt the ice under the runners, and allow the runners to go more smoothly over the ice.
24. Heavy people would put more pressure on the ice, causing more ice to melt, reducing the friction between the runners and the ice.
25. The circulation of the air removes the warm air from directly over your skin. Also, the air that is removed already has as much water as it can hold. When it is replaced by drier air, your body can continue to cool by evaporation.
26. For our comfort, we want to know how much of our perspiration will evaporate. The important thing is then how much of the air's capacity to hold water has already been used up--the relative humidity.

Chapter 12--How Heat Gets Around

1. At the higher temperature, there is a greater temperature difference between the inside of your house and the outside, so the heat flow rate is greater.
2. When the temperature is lower outside, the temperature difference between inside and outside is greater and the heat flow rate is greater.
3. The thermal conductivity of water is high, so if your clothes are wet, they will conduct heat away from your body faster than if they are dry.
4. The air gap provides a greater distance between the hot region and the cold region, so the heat flow is slowed down.
5. Things cook faster in water than in an oven. On a cold day wet clothing conducts heat away from your body faster than dry clothing.
6. No. Conduction relies on atoms or molecules running into each other. There are none in a vacuum.
7. The vacuum stops conduction, and the reflective surface stops radiation.
8. The hotter water has a lower density than the cooler

water, so it floats to the surface. We have tried to word the question so that "boiling" or "convection" are not acceptable answers.

9. The source of heat for the lower atmosphere is the ground. As you get farther from the ground, it gets colder.
10. As water vapor is carried aloft in convection currents, it reaches a height where it is cool enough for most of the water to condense, and this is the height at which a layer of clouds can form.
11. The ozone directly absorbs the ultraviolet radiation from the sun.
12. To promote cooling, you would want the roof to be reflective in the visible part of the spectrum and a good radiator in the infrared part of the spectrum. The opposite would promote heating.
13. At the ground.
14. The heat capacity of the air is great enough that it still takes a lot of heat to bring the air to its maximum temperature, and it takes a while for the air to absorb this heat.
15. (a) Conduction. (b) Conduction. (c) Radiation. (d) Radiation. (e) Conduction. (f) Conduction.
16. The distance between the very cold air and the warm water increases as the ice gets thicker, so the heat flow rate goes down. In a sense, the ice acts as an insulator.
17. Air. You want to fill the gap with the poorer conductor of heat.
18. As you break up the ice, the total amount of ice remains constant, but the surface area increases, so the heat flow into the ice increases.

Chapter 13--Waves

1. No. As the band marches, matter (the marchers) is getting from one place to another.
2. The spot moves up and down perpendicularly to the string, but not along the wave's direction of motion.
3. Highest pitched string would have the fastest waves. The wavelength is fixed by the length of the string. Since wavelength × frequency = speed, higher frequency

must mean higher speed.

4. The strings would have to be of different weights.
5. (1) Sound. (2) Waves in a spring.
6. It moves back and forth, along the direction in which the wave is moving.
7. 3 km (330 m/s × 9 s).
8. No. You won't hear it until the plane passes by, since it travels behind the plane.
9. If metals were perfectly rigid, sound would pass instantaneously from one end to the other, but we know that this is not the case.
10. 10^3 cycles/s = 10^3 Hz.
11. It would move back and forth, perpendicular to the direction in which the wave is moving, and it would go in simple harmonic motion.
12. If the point you pick is not at the end of the swing, the string will go through it twice in each cycle.
13. (a) (331 m/s)/(440 cps) = 0.75 m. (b) Half as great, 0.38 m.
14. (a) Speed stays the same. (b) Halve the wavelength.
15. The motions are back and forth, and there is no motion perpendicular to the direction of the wave.
16. (a) First. If you rotate the second, none will get through the first, so there will be no change. (b) As you start to rotate the filter, more and more will get through so the final signal increases. However, once it gets past 45 degrees you lose more at the second filter than you gained at the first.
17. None.
18. Sharp.
19. (a) 1.98 m. The change is in proportion to the wavelength. (b) 1.01 m and 2.02 m. The changes would be of the same amount, but with the opposite sign. (c) 0.98 m, 1.96 m. The changes would be twice as great as before.
20. This is when the sound is shifting from being sharp to being flat. Also, the pitch is changing the most quickly.

ANSWERS TO QUESTIONS

Chapter 14--Standing Waves

1. Yes.
2. Yes. The wave disturbance still oscillates at the point of constructive interference. It just means that the maximum value is the largest it can be.
3. Total destructive interference.
4. (a) Yes. (b) Yes, the string still goes up and down at that point.
5. The wavelength on the 440 string is twice that on the 880 string.
6. The string will appear reflected about the zero line.
7. The string is half a wavelength, or 2 m.
8. Change the tension of the skin.
9. One end fixed and the other free. The string is then only a quarter of a wavelength in the fundamental mode, so the wavelength is 4 times the length of the string. With both ends fixed, the wavelength is only twice the length of the string.
10. Rich and full, because various harmonics are being amplified in the throat, mouth, and other chest and oral cavities by resonance.
11. No. At a pressure node, the pressure is just equal to the average pressure.
12. No. Between the nodes, the pressure oscillates between being greater and being less than average. An antinode means that these oscillations go through the maximum pressure change.
13. The image of light and dark representing pressure variations will look the negative of the existing figure.
14. The open end is in contact with the surrounding air, so it must always be at the pressure of the surrounding air--a pressure node.
15. With a closed end, the wavelength in the fundamental is four times the length of the pipe. With an open end it is only two times the length. The closed end gives you the lower note.
16. You would hear a higher frequency tone that got louder and softer once per second (beats).

ANSWERS TO QUESTIONS

Chapter 15--Light--The Wave Side

1. Interference and diffraction.
2. The light is not coherent.
3. The light is not coherent.
4. The star that is 10 times as far away is 1/100 as bright in appearance.
5. It gets larger. The whole pattern spreads out.
6. They are the same. The pattern spacing only depends on the ratio of the wavelength and the spacing between the slits.
7. 1001 wavelengths.
8. The new diagram would show the short arrow representing the wave internally reflected at the top of the substance extended outward through the bottom. When this doubly reflected wave is out of phase with the directly transmitted wave, then destructive interference takes place.
9. Nonreflective coatings work when the spacing between layers in the coating, or the thickness of the coating, is just right to produce destructive interference in the reflected wave. This only works for a particular set of wavelengths.
10. We get a maximum reflection.
11. The wavelength of light is too small.
12. (a) Since the object is moving away from us, wavelength will increase by $(10^{-5})(700\ \text{nm}) = 7 \times 10^{-3}$ nm, to 700.007 nm. (b) The separation will increase by 1×10^{-3} nm to 100.001 nm. (c) The wavelengths would decrease instead of increasing.
13. The wavelengths of all lines increase in proportion to the rest wavelength, so the spacings increase.
14. No.

Chapter 16--Holography

1. A hologram is an interference pattern. If many wavelengths are present, they will not interfere destructively at the same point.
2. By splitting one beam, you get two beams whose phases are related.
3. Viewing.

4. In a hologram, the whole picture is contained in each piece of the hologram. Having both halves just improves the sharpness of the picture.

Chapter 17--Optics

1. We ignore diffraction and interference, but we use the fact that the waves change speed when they go from one medium to another.
2. 10 m (5 m to the mirror and 5 m back).
3. (a) $(3 \times 10^8\ \text{m/s})/(1.5 \times 10^8\ \text{m/s}) = 2$.
   (b) (600 nm)/2 = 300 nm.
4. (a) 1.2. (b) 1.2.
5. The same, since it is in air in both places.
6. Frequency does not change.
7. It doesn't.
8. Greater.
9. Gradual turns. In this way the light will always make glancing reflections off the edges, and will always be past the critical angle, so you always get total internal reflection.
10. Real.
11. Left and right are not really reversed in a reflection in a plane mirror. It is really front and back that are reversed, which is perfectly ok. We tend to define left-right with respect to up-down and front-back, so if front-back is reversed, we get a backward view of left-right. The situation is the same as when you face someone, in which case the other person's right is your left and left is your right, but that person's top is not your bottom.
12. (a) The parallel lines coming from the left will diverge upward to the left after reflection from the convex mirror. These diverging lines can be traced backward to the right (perhaps with dotted lines, to show that the light doesn't actually follow this path) until they come together at a point, a virtual image. (b) The image will be real, inverted, the same size as the original, and at the same distance from the mirror as the original (that is, the part of the object on the axis will be imaged on itself).
13. The rays will come off parallel to each other, but at

an angle with the axis.

14. When the object is less than one focal length from the mirror.
15. Virtual.
16. Blue, since the blue light rays will be bent more.
17. (a) They go through the focal point. (b) They bounce off making the same angle with the axis as when they came in, but on the other side of the axis.
18. (a) It goes through the focal point. (b) It passes through essentially undeflected.
19. Situation (a).
20. The density of the air changes with altitude, and as the density changes, the index of refraction changes. If we considered the atmosphere to be at constant density, then there would be a one-time bending when a light ray entered the atmosphere, but then the ray would continue straight without further bending.
21. White light has many different wavelengths, each of which would be focused at a different point, but all of the light, in focus and not, would expose the film.
22. The lens would not be symmetric about the axis of the lens.
23. Chromatic aberration.

Chapter 18--Optical Instruments

1. (a) Exposure times are the same since both cameras have the same focal ratio. (b) The longer focal length camera gives more detail.
2. Use the larger lens to get the brighter picture.
3. (a) The shorter focal length gives the brighter picture. (b) The longer focal length gives greater detail.
4. A long focal length lens will produce a large image. This is what you want if the object is very far away. A telephoto lens is constructed with several elements so as to give a large image while still being shorter than a simple lens that produced an image of the same size would be. (This definition can be found in many ordinary dictionaries. The point about wanting a large image will ordinarily be sufficient for an answer from students, although it does not meet the technical definition of telephoto.)

5. A series of lenses is necessary to correct for various aberrations. A series is also used to make camera lenses relatively compact compared to simple lenses.
6. (a) A small focal ratio means a relatively large diameter, which means that the lens must be good over this relatively large diameter. (b) The less of the lens you have to use, the more you will cut down on aberrations.
7. The image formation is essentially the same. The eye adjusts to different distances by changing the focal length of the lens. The camera adjusts by changing the distance between the lens and the film. The eye has a built-in exposure time, and the camera has an adjustable exposure time.
8. They refer to where the image falls with respect to the retina.
9. 1.8 m.
10. The diameter of the sun must be 400 times that of the moon.
11. Light of one wavelength.
12. The large objective telescope can gather more light.
13. Use a shorter focal length eyepiece.
14. The rays emerge from the eyepiece as parallel rays, just as if they were coming from a source "at infinity."
15. Reflecting telescopes are free of chromatic aberration. The mirrors can be made very large. The glass need not be perfect throughout; only its surface need be perfect. The mirror can be supported from behind, instead of just at the edges. However, a refractor need not have any of its light blocked by a mirror to bring the image out of the tube. Some of the advantages of refractors are relatively technical. For example, pinprick-size irregularities in the reflective coatings of mirrors are usually present, and scatter a small amount of light instead of focusing it. Thus for situations where an extraordinarily low amount of scattered light can be tolerated, such as for studies of the sun with coronagraphs, refractors are usually used. It is often also felt that a refractor gives better images than a reflector of the same size, something which amateur astronomers are often familiar.
16. As we decrease the focal length of the objective, the

image becomes larger; this property leads to the magnification being the focal length of the objective divided by the focal length of the eyepiece.

Chapter 19--Electricity at Rest

1. They are the same, since they both depend on distance in the same way.
2. We have an equal amount of positive and negative charge, so our net charge is zero.
3. Small number.
4. Charge is conserved. Before we had no net charge, and after we have two equal and opposite charges, so there is still no net charge.
5. It is not necessary. We need only observe particles whose charge changes by e or various multiples of e. Millikan's oil-drop experiment discovered e by searching for the least common denominator of his measured values.
6. The salts from your skin dissolve in the water, making it a good conductor.
7. Charge must be able to freely flow from one side of the object to the other.
8. Equal and opposite.
9. (a) 4 N. (b) 8 N.
10. (a) 100 N/C. (b) 200 N.
11. It will rotate until it lines up with the field.
12. The fact that it is neutral means that it can feel no net charge in a uniform electric field. However, since the charges are not in the same place, it can still feel a torque.
13. The two charges will not feel equal and opposite forces, so there will be a net force on the dipole. It will move toward the region of stronger field.
14. No.
15. It doubles.
16. The higher voltage picture tube will produce a faster moving beam of electrons and therefore a brighter image.
17. (a) The electric field increases. (b) The deflection of the beam increases.
18. Connect one lead to each plate controlling the up-down

motion of the beam. Increase the voltage between the plates controlling the left-right motion, steadily with time. The beam will trace out the voltage between the two connectors as a function of time.

Chapter 20--Electric Circuits

1. The electric field comes on in the wire almost instantaneously, so electrons start to flow everywhere in the wire.
2. Right to left, opposite to the direction of the negative charge flow.
3. Left to right, in the direction of positive charge flow.
4. Yes. Each electron and proton contributes to a current flow from left to right.
5. No. The wire is neutral.
6. (a) Increase resistance. (b) Decrease the resistance. (c) No change.
7. (a) Potential difference is the same. (b) The lower resistance resistor will have the greater current.
8. Both decrease the resistance, because they provide a wider path for the current flow.
9. Closed.
10. In each case we are doubling the distance that the current must flow through a particular type of wire, so we are doubling the resistance.
11. (a) Yes. (b) No.
12. No. As long as it is still on the same circuit, then it will draw the same amount of current through that circuit and the fuse will blow.
13. All of the appliances are in parallel. A short circuit will provide one zero-resistance path for the current, reducing the total resistance of the whole circuit to zero.
14. Desirable: Toaster, electric heater, light bulb. Undesirable: Heating in wires producing fire, heat generated in computers, the burning out of a light bulb.
15. The fuse blew because the student was trying to pass too much current through the ammeter. By using a larger fuse this current will flow through the ammeter, and burn it out (assuming that the safety margin over

the fuse's limit is passed).

16. The 100-watt bulb has twice the resistance.
17. Calculators, computers, radios, digital watches, factory controls, some new auto engine controllers.
18. The circuit is etched into the microcircuit. The standard circuit has individual components with wires going from component to component.
19. The water, with some salts from the skin dissolved in it, is a very good conductor and lowers the resistance of the body.
20. Burns and damage to the nervous system.
21. The wire has a very low resistance over that small length, so the potential difference between the bird's feet is not very large. Further, the bird has a high resistance. As a result, very little current flows through the bird.

Chapter 21--Magnetism

1. On the smallest scale, magnetism arises from currents flowing within atoms, creating magnetic moments. In a bar magnet, these currents work together to produce a large magnetic field.
2. The one moving at 10 m/s perpendicular to the field. The other feels no force.
3. (a) The speed of the electron doesn't change. (b) The energy of the electron is constant, so no work is being done by the field.
4. They can be trapped by a magnetic field.
5. The particle spirals around field lines, getting closer to a pole, and slowing down, until it comes to a rest, and then starts to spiral back in the other direction.
6. Yes. As the particle is being slowed down, its kinetic energy is decreasing, so the field is doing work on the particle.
7. A plasma is too hot to hold in a normal bottle, but might be held in a magnetic bottle.
8. First the molecules must be ionized, so they have a net charge, and can therefore feel a magnetic force. If they are then accelerated to the same energy and pass into a region of uniform magnetic field, they will go in circular orbits, but because the masses are

different the radii of the orbits will be different, and the two beams can be separated.

9. The current loop will feel a torque, due to the magnetic forces on each part of the loop. The loop will rotate until its magnetic moment lines up with the magnetic field.
10. You could determine magnetic north using a compass, and true north by the stars.
11. The metal in the building can alter the magnetic field pattern inside the building.
12. Both the compass needle and the filings feel the torque produced by the earth's magnetic field. However, the filings are not free to turn, because of the friction between them and the paper, while the pivot of the compass provides very little friction, so the needle is free to turn, and lines up with the magnetic field.
13. There is no single "monopole" of magnetic charge, analogous to the electric charge on the electron and proton. All magnetic fields arise from currents, and the most basic field that a current can produce is a dipole field.
14. If we have the north pole of one magnet pointing toward the south pole of the other magnet, then the magnetic moments of the two magnets are pointed in the same direction. This means that the currents in the magnets are circulating in the same direction, and the magnets are attracted to each other.
15. The magnetic moments of electrons orbiting in opposite directions cancel in most atoms.
16. If some of the metal is ferromagnetic, it can become magnetized in a magnetic field, and the watch will then experience magnetic forces that can interfere with the precise movements of the timing mechanism.
17. The dropping may provide enough energy to throw the magnetic domains out of alignment, and weaken the magnetic field.
18. Jupiter was a strong source of radio noise, and it was thought that this comes from belts of charged particles trapped in a magnetic field, like in the earth's radiation belts. This deduction has now been verified.
19. We had thought that planets need a molten core to have a magnetic field, and Mercury does not seem massive enough to have a molten core. We had also thought that

rapid rotation would be necessary, although Mercury rotates relatively slowly.

20. Sunspots.
21. It can affect the weather, radio communications, and produce auroras.

Chapter 22--Changing Fields: Motors to Maxwell

1. A changing magnetic field can produce an electric field. Another way of looking at it is to say that a changing magnetic flux through a loop can cause an EMF around the loop.
2. (a) The magnetic field is getting stronger as the magnet gets closer. (b) You are changing the direction of the magnetic field, as well as the strength of that part of the field that passes through the loop. (c) This increases the magnetic field strength. (d) This decreases the magnetic field produced by the second loop.
3. AC is the natural form to generate. It is easily transformed in voltage, allowing it to be transmitted at a high voltage and used at a low voltage.
4. (a) Electric clocks will run slow, since they keep time by counting cycles on the AC line. (b) At the end of the day, the company can send out the proper number of extra pulses, to make up for those that are missed during the day.
5. As you turn the crank faster, you are increasing the rate at which the magnetic flux through the coil changes.
6. Step up--in automobile engines, to get the high voltage needed for the spark plugs. Step down--in a toy train set, to convert the line voltage to a low safe voltage for running trains.
7. Yes. They are the same. They are just hooked up in reverse from each other.
8. In a generator, we turn a crank, doing work, to produce a changing flux through the coil, and therefore generate electric current. In a motor, we do the opposite. The changing current causes a changing flux in the coil, causing a changing torque, and the coil rotates, so we get work out.
9. The sudden change in current causes a sudden change in

the flux through the inductor, and the inductor will oppose this by producing a very large voltage to try to keep the current flowing. This voltage may be large enough for sparks to jump across the opening in the switch.

10. The r.m.s. voltage is directly related to the average power that you can get out, and this tells you how much work you can do over a given time.
11. It is an oscillating system and the energy oscillates back and forth between two forms, electric and magnetic (instead of kinetic and potential for the pendulum or spring).
12. All waves that had been known until that time needed a medium to travel through.
13. (a) They travel at the same speed. They are oscillating electric and magnetic fields, as described by Maxwell's equations. (b) They have different wavelengths (frequencies).
14. The electric and magnetic fields will oscillate as the pulse passes by.
15. No. A changing magnetic field would produce a changing electric field, but then the changing electric field would do nothing, so any wave would just stop.
16. He recognized the symmetry between electric and magnetic fields, and included the fact that a changing electric field can produce a magnetic field.
17. In this case, we have to define range by the number of decades included, and this would be the infrared.

Chapter 23--Quantum Theory and Atomic Structure

1. The quantization of light. The wave nature of particles. The fact that atoms can only exist in particular states. The existence of antimatter.
2. It didn't properly predict the high-frequency part of the spectrum (the "ultraviolet catastrophe").
3. Wave: Interference, diffraction, refraction with slowing down in media with higher index of refraction.
   Particle: Black-body radiation spectrum, photoelectric effect, emission of specific energies by atoms.
4. Classical theory of black-body radiation predicted the the amount of energy given off by black bodies gets

larger and larger as you go to shorter wavelengths, with an infinite amount of energy being given off.

5. Ultraviolet. It has a shorter wavelength (higher frequency), and energy = frequency × Planck's constant.
6. Stronger light means more photons, which means more electrons, not more energetic electrons.
7. It was an empirical formula, not derived from any theory.
8. Most of the alpha particles would have been scattered, but only very slightly.
9. (a) He assumed that electrons can circle the nucleus in certain stationary states, and that when an electron went from one state to another, it had to emit or absorb a photon with the right energy, and that the angular momentum of the electrons in their orbits was quantized. (b) He was able to explain the spectrum of the hydrogen atom.
10. (a) Emission of individual electrons in the photoelectric effect. Particle-like trajectories of electron beams. (b) Electron diffraction. The existence of stationary states within an atom.
11. The slower one (since its energy is lower).
12. The probability of finding the electron in a particular place.
13. The car is so massive that the uncertainty in momentum is much less than the momentum of the car, and the uncertainty in position is much less than the size of the car.
14. The uncertainty principle, the fact that we can't predict exactly where a particular particle will be at all times. Hidden-variables theories say that the problem is not with the electron and other particles, but with the measuring processes.
15. Since only two electrons can stay in a given level, it means that higher levels will be more populated than if the exclusion principle did not apply.
16. In some types of experiments matter behaves like a set of particles, and in other types of experiments it behaves like a collection of waves.
17. (a) The wavelength becomes shorter. (b) The total amount of energy increases.
18. (a) 1000 K. (b) 1000 K. (The hotter one gives off more energy at all wavelengths, even though the peak of

its spectrum comes at a shorter wavelength.)

19. The sun's light is in the visible part of the spectrum, since the sun is about 6000 K, and the ground absorbs this radiation (most of which can pass through the earth's atmosphere). However, the earth is only heated to about 300 K, so the radiation given off by the earth is mostly in the infrared, and the earth's atmosphere absorbs most of the infrared radiation.
20. It is halved.
21. Hotter. As you heat something, the peak of the spectrum shifts to shorter and shorter wavelengths. For the star, the blue color indicates a shorter wavelength peak than the yellow color of the sun.
22. (a) Double the energy. (b) Halve the energy.
23. In both cases, there are experiments in which we see particle behavior and there are experiments in which we see wave behavior.
24. Einstein used the fact that radiation comes in bundles of energy (h × frequency).
25. The energy of the light that is ejecting the electrons (and the type of metal used).
26. The intensity of the light.
27. Classical physics predicted that atoms should emit light at all wavelengths, and that the electrons should spiral into the nucleus, making atoms unstable.
28. (a) He was able to explain the particular wavelengths. (b) He was not able to explain why some lines are stronger than others.
29. He had to assume that stationary states existed, that transitions could only be accompanied by the emission or absorption of a photon, and that the angular momentum of the states is quantized.
30. The wavelengths are so small.
31. Deciding how much time to allow to get to work. A baseball manager deciding which batter to send up. Evaluating risks in deciding whether to do something.
32. The one in the breadbox, since it has a smaller uncertainty in position.
33. Chemical reactions involve the reactions between atoms, in which the outermost electrons are involved. In this sense, we can apply the physics we know to predict, in principle, how these electrons will act and how the atoms will combine. Certain kinds of chemistry (such

as those using large organic molecules), though, deal with situations too complicated to be able to solve with straightforward applications of laws of physics.

34. The orbits were extended to elliptical orbits. A wave mechanics description of why the electrons can only stay in certain states was proposed independently by Schrödinger and Heisenberg. Quantum theory was put in a proper, relativistic language, and the prediction of antiparticles came out of this. The quantum mechanical theory of the electromagnetic interaction--QED--was proposed.
35. A virtual photon is the carrier of the electromagnetic force. When two particles interact via the electromagnetic force, one or more virtual photons pass between the interacting particles.

Chapter 24--Lasers

1. A laser gives off radiation in the visible part of the spectrum and a maser gives off radiation in the microwave part of the spectrum. The basic process is the same, however.
2. An electron in the atom is in a higher energy state than normal.
3. Two.
4. The same.
5. The photon emitted in stimulated emission is in phase with the incoming photon.
6. There are many atoms in the excited state, each of which can be stimulated to emit radiation at the same frequency.
7. The mirrors allow the radiation to pass through the material many times, providing more stimulated emission than if the radiation went through only once. The effective size of the cavity is increased.
8. If they were both 100% reflecting, no radiation would ever get out. Usually only one of the mirrors is partially transmitting.
9. Cutting and welding in industry. Surgical uses. Lining up a telescope or railroad system. Making a hologram.
10. A large number of laser beams might be used to compress a fuel pellet enough to start fusion.

ANSWERS TO QUESTIONS

Chapter 25--Solid-State Physics

1. A crystal has a regular lattice structure, while an amorphous solid does not have such a regular structure.
2. Amorphous solid: Glass, ceramics.
   Crystal: Salt (sodium chloride), diamond.
3. They have filled valence shells so they will neither give up nor receive electrons.
4. When there are very few electrons in the valence shell, atoms will give up those electrons fairly easily, to be left with a filled shell. Therefore, in a solid, these electrons are available to conduct electricity.
5. It would collapse.
6. They are very similar, arising because electrons are only in certain energy states. The bands in solids are many levels grouped together.
7. The valence band is the highest band in which electrons are normally found, and the conduction band is the lowest band in which the electrons are free enough to conduct electricity. If the two bands are the same, or are close together, then the solid can conduct electricity.
8. If there is a large gap between the valence band and the conduction band, then the solid will be a good insulator. If there is no gap, then it will be a good conductor.
9. The electrons in the metals might act as the means for transmitting the heat.
10. A small gap between the conduction band and the valence band.
11. (a) Right to left. (b) Right to left.
12. It provides more free electrons to conduct electricity.
13. (a) It conducts normally. (b) It will not conduct. (c) It will conduct.
14. They work in reverse from each other. In an LED, we provide a voltage and a photon is given off. In a photodiode, a photon is absorbed and we get a small voltage.
15. Transistors are smaller, require less power to run, don't burn out, and don't heat up as much as tubes.
16. Their small size means that crucial parts can be closer together. Also, they consume less power and heat up less, so not as much air conditioning is required.

17. The MOSFETs are smaller, and circuits can be made more compact. They are also suited for manufacture right into the printed circuits. MOSFETs can take relatively high power levels without distortion.
18. The magnetic field cannot penetrate the superconductor.
19. Materials have to be cooled to very low temperatures before they become superconductors.

Chapter 26--Inside the Nucleus

1. About the size of the tip of a pin.
2. The positively charged protons repel each other.
3. The neutrons are unstable and will decay.
4. It provides the strong force to bind the nucleus, without providing any electric repulsion.
5. (a) i and ii. (b) i and ii. (c) ii and iii. (d) ii and iii.
6. Since an atom is neutral, it will have the same number of electrons orbiting the nucleus as it has protons in the nucleus. It is the electrons orbiting the nucleus that determine the chemical properties of the element.
7. (a) The same. (b) There is no electric force between the neutrons (although there is a small magnetic force), so the electromagnetic force between the two protons is greater.
8. The nuclear force holds it together, and the electric force pushes it apart.
9. A force in which the strength of the force falls off so quickly that it has no effect outside a very small range.
10. It is the intermediary that transmits the force from one nuclear particle to another.
11. They must go into the allowed energy levels, and they must obey the exclusion principle.
12. It has been successful in explaining which nuclei are stable.
13. It has been useful in explaining fission.
14. The two models are really discussing different aspects of nuclear structure, and both are needed to describe a nucleus.
15. More stable. More energy will be needed to break it apart.

16. More stable.
17. (a) mgh/c^2:
$$\frac{(75\ \text{kg})(9.8\ \text{m/s}^2)(10\ \text{m})}{(3 \times 10^8\ \text{m/s})^2} = 8 \times 10^{-14}\ \text{kg}.$$
(b) $(8 \times 10^{-14}\ \text{kg})/(1.7 \times 10^{-27}\ \text{kg}) = 5 \times 10^{13}$.
18. Alpha: Final charge 2 less than original, final mass number 4 less than original, and final atomic number 2 less than original.
Beta: Final charge can be 1 more or less, depending on whether a positron or electron is given off. Final mass number is the same, and final atomic number can be one more or less, depending on which is given off.
Gamma: None is changed.
19. (a) Alpha. (b) Alpha. (c) Gamma.
20. A.
21. A particle can pass through a region for which it really does not have enough energy to enter. The alpha particle can leave the nucleus in this way.
22. It speeds up since it is repelled by the positive charge of the nucleus.
23. Beta.
24. Its presence was needed to account for missing energy and angular momentum in beta decays.
25. They interact with matter only by the weak interaction.
26. If they were buried at some time, so they could no longer get the carbon-14 (^{14}C) from the atmosphere.
27. The processes are similar, although in the nuclear process higher energies are involved. Gamma rays are just high-energy photons.
28. It must have enough energy to overcome the electric repulsion of the nucleus.
29. They can be accelerated electrically, as in accelerators, or they can be heated to high temperatures, as in the center of the sun.
30. The binding energy per nucleon is greatest for moderate mass nuclei (iron-56 = ^{56}Fe), so most reactions that bring you toward iron-56 tend to give off energy.
31. There must be a high enough concentration of nuclei that can absorb a neutron and undergo fission.
32. The energy available in 0.007 of the sun's mass is $0.007 \times mc^2$ =
$$(2 \times 10^{30}\ \text{kg})(0.007)(3 \times 10^8\ \text{m/s})^2 = 1.3 \times 10^{45}\ \text{J}.$$

This will last for

$$(1.3 \times 10^{45}\ \mathrm{J})/(4 \times 10^{26}\ \mathrm{J/s}) = 3 \times 10^{18}\ \mathrm{s} = 10^{11}\ \mathrm{yr}.$$

This is how long the sun could last if it could use all of its mass for fuel. We think that the sun is now half as old as this maximum.

Chapter 27--Energy for Tomorrow

1. Identifying possible sources of energy, for example, possible nuclear reactions, or various ways to use solar energy. Searching for possible side effects of various methods that have been proposed.
2. They are both remnants of of the same types of marine life, subjected to the same conditions.
3. (a) Explosions in transportation, and carbon dioxide released into the atmosphere in burning. (b) Spills in transportation and drilling, and pollution in burning. (c) Dangers in mining, environmental problems in strip mining, pollution from burning high-sulfur-content fuel, and carbon dioxide released into atmosphere in burning. (d) Radioactive leaks in operation of reactor, meltdown causing sudden release of radioactivity, problems in storing radioactive wastes.
4. Uranium in which the amount of ^{235}U is increased over its natural abundance.
5. Most uranium is ^{238}U, which will not undergo fission, and will absorb neutrons. Enough ^{235}U must be present to keep the chain reaction going.
6. The centrifuge creates an artificial gravity (a pseudo-force) separating the heavier and lighter gases.
7. Gases are necessary in the various separation processes, and UF_6 is a gas.
8. It slows down the neutrons, so they can be absorbed more easily.
9. (a) A water molecule in which one hydrogen atom has been replaced by a deuterium atom. (b) Heavy water absorbs fewer neutrons than regular water.
10. The fuel is not enriched enough.
11. They absorb neutrons to control the rate at which nuclear reactions take place.
12. Once the heat is generated, either by burning fuel or

by nuclear reactions, the heat is used to generate steam, which then turns a turbine, which turns a generator shaft.

13. Each fission gives off extra neutrons, which can be used to make more fuel.
14. Alpha--helium nucleus; beta--electron or positron; gamma--photon.
15. An energetic particle--alpha, beta, or gamma--strikes the material and is absorbed, or at least scattered, leaving some or all of its energy in the material.
16. It can ionize material.
17. Somatic effects affect only the exposed person and genetic effects affect the genetic material and thus future generations.
18. The presence of salt means that no water has been flowing through for a long time.
19. The fuel is very plentiful, and there are schemes for which the radioactive wastes will be at a much lower level than in fission reactors.
20. Generally, the reactions that give off energy are those for which the final products are closer to iron-56 (^{56}Fe) than the initial reactants.
21. (a) A magnetic field arrangement for containing a plasma, or gas of charged particles. (b) The magnetic field lines come close together at the ends (meaning that the field is stronger at the ends). As the particles go back and forth, they are slowed down and turned around at the ends.
22. For nuclear fission to take place, the fuel must be brought to a very high temperature to overcome the electric repulsion of the nuclei. At the high temperatures necessary, the material becomes ionized.
23. Estimate that a normal household requires about 5000 W. This would mean 50 m^2.
24. The price of converting solar energy into electricity must be improved greatly, which involves finding more efficient processes of making the conversion and ways of making the materials involved at greatly reduced prices.
25. It is present, whether we use it or not. It looks like it will only be able to provide 10% of our needs under the best conditions. It is most readily available only in certain locations.

ANSWERS TO QUESTIONS

Chapter 28--Quarks

1. gravitational--graviton
   weak--W
   electromagnetic--photon
   strong--pi meson.
2. photon, pi meson.
3. A large amount of energy is needed to create these particles, and the accelerators raise the energy of the colliding particles.
4. From the kinetic and rest energy of the colliding particles.
5. Bubble chambers--particles leave tracks that are photographed. Spark chambers--particles pass through and spark and jump to mark their path.
6. We know the energy, momentum, angular momentum, etc., of particles going in and of the particles that we can detect coming out, so we use conservation laws to deduce the properties of particles we cannot see.
7. When the energy is right for the creation of a new particle, the number of reactions will rise sharply. This is called a "resonance."
8. (a) Lepton. (b) Lepton. (c) Hadron. (d) Hadron. (e) Hadron. (f) Neither.
9. The electron and the muon are known to have their own neutrinos, and since the tau is the next in this series, we believe it should have its own neutrino. Some direct evidence is now accumulating.
10. Hadrons include baryons and mesons.
11. The discussion will include material on pages 478-479 and perhaps also Occam's razor (page 111) from the discussion of the Copernican theory.
12. For each symmetry, there is a conservation law.
13. Hint: Each symbol is a number (1, 2, 3, 4, 5) and its mirror image.
14. Charge would not be conserved.
15. There must be some conservation law that we don't know about, which would be violated if the reaction took place.
16. The strong force honors more conservation laws than the weak force.
17. Energy, momentum, angular momentum, charge, parity (not in weak interactions), strangeness (not in weak

interactions).

18. It showed that certain symmetries in nature are not perfect.
19. Experiments show that protons have concentrations of matter in them ("partons"). The theory is successful at predicting particles that exist and their properties.
20. The allowed combinations are:

uuu	charge = 2	$\bar{u}\bar{u}\bar{u}$	charge = −2
uud	charge = 1	$\bar{u}\bar{u}\bar{d}$	charge = −1
udd	charge = 0	$\bar{u}\bar{d}\bar{d}$	charge = 0
ddd	charge = −1	$\bar{d}\bar{d}\bar{d}$	charge = 1

21. $u\bar{u}$--charge 0; $u\bar{d}$--charge 1; $d\bar{d}$--charge 0; $s\bar{s}$--charge 0.
22. Color, like spin orientation, is a property that can be changed for a particle. Also, two quarks of the same color cannot be in exactly the same state, but two quarks of different colors can.
23. Colors are related to how the quarks combine. Flavors are the basic properties of matter that are carried by the quarks.
24. They can change.
25. According to one theory, the force between two quarks doesn't fall off with distance. So much energy is needed to pull two quarks apart that new quarks and antiquarks are created in the process, so we never end up with a free quark (string model and bag model).
26. It pointed to a new property of matter (charm) and thus to the existence of a new quark.
27. A new particle, the upsilon, is believed to be the combination of this fifth quark and its antiquark.
28. Mostly on the basis of theoretical predictions concerning the idea that the number of quarks may match the number of leptons. Since the leptons seem to come in pairs (each lepton with its own neutrino), the quarks thus probably come in pairs too.
29. Just two, u and d.
30. More of the energy is available to create new particles, since the total momentum of the two beams is zero, and all of the energy can be used.
31. Answer may be satisfied with the old "fundamental particles" (protons, electrons, mesons, etc.) as a defined term, may use quarks as fundamental particles,

or may say that there are probably particles that make up each level and thus there are no really fundamental particles.

Chapter 29--Astrophysics

1. Gas pressure from the energy from fusion at the star's core.
2. Hydrogen is being converted into helium.
3. Not as many solar neutrinos as predicted have been observed, and it may be that our theories of what goes on inside stars are not totally correct or else that we don't know as much about neutrinos as we had thought. (The latter point is connected with the new experiments that indicate that neutrinos may have mass; these experiments need confirmation before being accepted, though.)
4. Neutrinos interact weakly with matter, and most pass right through the earth. Only with a large amount of detector material will enough reactions take place to be detected.
5. Electron degeneracy pressure.
6. Iron-56 is the most stable nucleus, so when it forms, none of the nuclear reactions that take place gives off energy.
7. Neutron degeneracy pressure.
8. Nothing; it becomes a black hole.
9. The visible star has an unseen companion whose mass is deduced to be greater than that required for a black hole, and which we know cannot be supported by ordinary gas pressure because it is not a main-sequence star (since a main-sequence star of that mass would be bright enough for us to see). The idea that a black hole is present is made plausible by the detection of x-rays from the system, presumably from the accretion disk composed of particles being drawn into the black hole. The presence of the x-rays, although it caused our attention to be drawn to this system, is not by itself enough to prove that a black hole is present.
10. They did not have sufficient mass to heat up enough for nuclear reactions to take place.
11. There is some evidence for invisible companions around

some stars in the wobbling motion of these stars as they move across the sky and in varying Doppler shifts.

12. (a) There was no means of determining their distances. (b) Cepheid variables were observed in some of these nebulae, and the relationship between their periods of variability and brightness allowed them to be used as distance indicators.
13. We measure the recession of distant galaxies from their Doppler shifts, given that we have methods of measuring the distances to these galaxies (from measurements of Cepheid variables, for example, although other methods are used for the farthest galaxies).
14. If they are as far away as their redshifts indicate, then they give off a large amount of energy, but the energy is created in too small a volume of space for normal processes of energy generation to be sufficient. Most astronomers now believe that the presence of giant black holes in the cores of quasars explains where quasars get their energy, solving the energy problem.
15. Expansion of the universe, cosmic background radiation, the existence of elements that were probably created in the big bang.
16. An open universe will expand forever. In a closed universe, the expansion wll stop and the universe will eventually collapse. It turns out that an open universe is infinite in extent while a closed universe is finite.
17. Comparing the Hubble constant for nearby and distant galaxies, determining the relative amount of deuterium in the universe.

SAMPLE QUIZZES AND EXAMINATIONS

QUIZ ON MECHANICS--CHAPTERS 1-6

1. What information do you need to predict the position of an object at all times?

Answer: You have to know the position and velocity at some time. You must also know the force at all times, since this gives you the acceleration at all times.

2. Draw a graph showing the position of an object at various times if that object moves at a constant speed. On the same graph, show the position of another object, starting at the same place at the same time, but moving faster.

3. You are sitting in a chair. Draw a diagram showing the forces acting on you.

4. You go from the earth to the moon, where the acceleration of gravity is 1/6 that on earth. (a) How does your mass change? (b) How does your weight change?

Answer: (a) Your mass stays the same. (b) Your weight drops by a factor of 6.

5. Two ice skaters of equal mass head for each other. They travel at the same speed in opposite directions. When they get close enough, they hold onto each other, so they can't fly apart. Describe the motion of the two skaters, after they join together. Explain your answer in terms of conservation of momentum.

Answer: The two skaters start with equal and opposite momenta, so their total momentum is zero. After the collision their momentum must still be zero, by conservation of momentum. Since they are stuck together, the only way for their momentum to be zero is for both of them to be at rest, so they don't go anywhere.

6. You carry a ball up a ladder, doing a certain amount of work. You drop the ball. When the ball reaches the bottom of the ladder, how does its kinetic energy compare with the amount of work you did in carrying the ball up the ladder? (Ignore air resistance.)

Answer: Gravity is a conservative force. When you bring the ball to the top of the ladder you increase its potential energy by the amount of work done. When the ball reaches the bottom of the ladder, all of this additional potential energy is converted into kinetic energy, so the kinetic energy equals the work done.

7. An ice skater starts whirling around with her arms

spread out. As she brings her arms in, describe her motion, and the reason for it.

Answer: As she brings her arms in, her moment of inertia decreases. Since her angular momentum must remain constant, by conservation of angular momentum, her angular speed must increase, so she spins faster.

QUIZ ON RELATIVITY (CHAPTERS 7–9)

1. (a) What led Newton to believe that there must be some force acting on the moon? (b) What led Newton to deduce the inverse square law for gravitation? (c) How did Newton demonstrate that the inverse square law for gravitation was correct?

Answers: (a) Since the moon is moving in a curved (almost circular) path, the moon must be accelerating. According to Newton's second law of motion, each acceleration must be caused by a force. (b) Newton was guided by Kepler's laws of planetary motion. (c) Newton showed that the acceleration of an apple on earth and the acceleration of the moon are in the proportion that is predicted by the inverse square law.

2. Considering the effects of the moon only, explain why there are two high tides a day.

Answer: As you get farther from the moon, its gravitational pull gets weaker. The pull on water on the side of the earth closest to the moon is greater than the pull on the center of the earth, so this water is pulled away from the earth, creating a high tide. The pull of the moon on the center of the earth is greater than the pull of the moon on the water on the far side of the earth. This pulls the earth away from the water on the far side, creating the second high tide.

3. What is the significance of the Michelson–Morley experiment?

Answer: It showed that the earth cannot be moving through the ether. Together with the observation of the aberration of starlight, this showed that the ether does not exist, and that light can travel through a vacuum.

4. What is time dilation?

Answer: If one observer is moving along with a clock and another is moving with respect to the clock, the observer moving with respect to the clock will see the clock run slower than the observer moving with the clock.

5. Describe two observational tests of general relativity.

Answer: 1--Bending of starlight by the sun. As starlight passes close to the sun, the path of the light is bent slightly around the sun, making the star appear slightly farther from the sun. This has been confirmed during solar eclipses. 2--Advancement of the perihelion of Mercury. When all known effects are taken into account (not including general relativity), there is 43 arc seconds per century advancement in the perihelion of Mercury's orbit that cannot be explained, but can be explained by general relativity.

QUIZ ON FLUIDS AND HEAT (CHAPTERS 10–12)

1. (a) Why does the pressure increase as you go deeper in the ocean? (b) What determines whether a boat will sink or float?

Answers: (a) Each layer of water must be able to support the weight of the layers above, so as you go deeper, there is more weight to support, and a larger pressure is needed to support it. (b) The buoyant force, which is the difference between the upward force on the bottom of the boat and the downward force on the top of the boat, must exceed the weight of the boat in order for the boat to float. This happens if the boat weighs less than the water it displaces. Another way of saying this is that the boat will float if the average density of the boat is less than the density of water.

2. In general, how do we measure temperature?

Answer: We measure a property of some object that changes with temperature, such as the length of a column of mercury in a tube.

3. What types of processes are forbidden by the second law of thermodynamics?

Answer: The second law says that we cannot have an ideal refrigerator, that is, a refrigerator that transfers heat from a cold object to a hot object without our providing any work. Also, we can't have a perfect engine, that is, an engine that absorbs heat and converts it all into work, with no heat exhausted.

4. Briefly describe the three mechanisms by which heat gets

around.

Answer: Conduction--Energy is transmitted from atoms or molecules in a substance to their neighbors via collisions. Convection--Matter actually moves along with the heat. Radiation--Energy is transmitted by light or infrared radiation, or any other form of electromagnetic radiation.

5. How does the greenhouse effect work?

Answer: Visible radiation from the sun passes through the atmosphere and is absorbed by the ground. The ground radiates energy in the infrared, which is absorbed by the atmosphere, heating the atmosphere.

QUIZ ON WAVES AND OPTICS (CHAPTERS 13-18)

1. Sketch the appearance of a string with a harmonic wave moving down the string. Indicate the amplitude and wavelength on the sketch.
2. (a) What is the difference between a longitudinal wave and a transverse wave? (b) Which one can be polarized?

Answers: (a) In a transverse wave, individual particles in the medium through which the wave is traveling move perpendicular to the direction of motion of the wave. In a longitudinal wave, the individual particles move back and forth along the direction of motion of the wave. (b) Transverse.

3. A string is tied down at both ends. For what wave-lengths are standing waves possible on the string? Express your answer in terms of the length of the string.

Answer: For the fundamental mode, the wavelength is twice the length of the string. For the first harmonic, it is the length of the string, then 1/2 the length, 1/4 the length, and so on.

4. In Young's double slit experiment, what determines whether a point on the screen will be light or dark?

Answer: If the light beams from the two slits arrive in phase, there will be a bright spot. If they arrive out of phase, there will be a dark spot.

5. An object is placed in between 1 and 2 focal lengths distance from a thin converging lens. Draw a diagram showing how to locate the image of this object.
6. Compare the manner in which a camera and the eye adjust

to objects at different distances.

Answer: For the camera, the focal length of the lens is fixed, so the distance between the lens and the film is varied. For the eye, the distance between the lens and the retina is fixed, so the focal length of the lens is adjusted by pressure from the surrounding muscles.

QUIZ ON ELECTRICITY & MAGNETISM (CHAPTERS 19-22)

1. If there are so many charged particles around, why aren't you held into or thrown out of your chair by the electrical forces?

Answer: Matter is essentially neutral. There is as much positive charge as there is negative charge, so the net electrical force between two large objects can be zero.

2. Sketch the electric field lines around a positive charge. Show, on your sketch, the direction of the force that an electron would feel when placed near the charge.

3. You move a positively charged ball from point A to point B, in a direction opposite to the electric field. You do a certain amount of work in this process. You then let go and the ball accelerates back toward point A. When it gets to point A, how does its kinetic energy compare with the work you did in moving it from A to B?

Answer: The kinetic energy will be the same as the work, since the electrical force is conservative.

4. Why do wires with currents flowing get hot?

Answer: The electrons are accelerated by the electric field, but they quickly bump into the stationary positive ions, and lose their energy to the ions, heating the wire.

5. Describe what happens to a magnetic dipole placed in a magnetic field.

Answer: The dipole will rotate until the dipole moment is lined up with the magnetic field.

6. Why is it dangerous to open the switch in a circuit with an inductor when a current is flowing?

Answer: The inductor creates an emf that opposes any change in the current in the circuit. When the switch is opened, the current will drop quickly and a very large emf will be induced. The emf will be great enough to make sparks jump across the switch.

SAMPLE QUIZZES

QUIZ ON MODERN PHYSICS (CHAPTERS 23–29)

1. What is the evidence that light behaves as a particle?

Answer: 1--Blackbody radiation. To produce a theoretical blackbody spectrum that matched the observed ones, Planck had to assume that the energy came in discrete bundles. 2--Photoelectric effect. To explain the observations of the ejection of electrons from a metal on which light is shining, Einstein proposed the existence of photons, with each photon transfering its energy to one electron. 3--Compton scattering. In the scattering of light and electrons, the light behaves like a particle.

2. What is the physical meaning of the wave function of an electron?

Answer: It gives the probability of finding the electron in a particular place.

3. How does the band structure of an insulator differ from that of a semiconductor?

Answer: In a semiconductor the energy gap between the valence band and the conduction band is very small. In an insulator, it is large.

4. (a) What is the purpose of the neutron in nuclear structure? (b) Why can't we add a large number of neutrons to a given nucleus?

Answers: (a) The neutron provides the attractive strong nuclear force without the repulsive electric force, and thus helps hold the nucleus together. (b) The neutron is unstable to beta decay. If there is a large excess of neutrons in a nucleus, some will become protons.

5. (a) What are the advantages of nuclear fusion as an energy source? (b) Why don't we use it now?

Answers: (a) The fuel is very plentiful and it does not appear that fusion reactors will have all of the problems involved with radioactivity that fission reactors have. (b) Very high temperatures are required to get fusion started because the nuclei repel each other, and we don't yet know how to control the fuel at such high temperatures (10 million K).

6. Why do we believe the quarks exist, even though a free quark has never been seen?

Answer: Theories involving quarks have explained a large body of experimental data. Allowed combinations of quarks explain all of the known hadrons, and no hadrons are predicted to exist which could have been found and were not. The theory may even tell us why individual quarks cannot be observed. Experimentally, we can see that the charge in a proton is concentrated into three clumps, as predicted by the quark theory.

FIRST HOUR EXAM

Do all questions. Show your thinking so that we can give part credit when appropriate. Write in blue or black ink.

1. Describe the laser fusion experiments now going on. Briefly discuss 6 different fields of physics that enter in. What are some advantages of fusion over other methods of supplying energy?

2. You are driving along a 2-km stretch of highway. For the first km you travel at 30 km/h. You would like to travel fast enough during the second hour so that your average speed for the whole 2 km is 60 km/h. Can you do it? (Hint: Think in terms of the time it takes.) How fast will you have to go?

3. The acceleration of gravity on the moon is one-sixth that on the earth. Make a table to show how far an object would fall on the moon after 1, 2, 3, 4, 5, 6 seconds.

4. Suppose a spring is hanging from the ceiling. You weigh 750 N and hang from the bottom of the spring. (a) What force does the top of the spring exert on the ceiling? (b) What force does the ceiling exert on the top of the spring? (c) Draw a diagram showing the forces acting on the spring.

5. A pair of ice skaters performs a trick in which they skate toward each other, join hands, and spin around. If one skater has a mass of 50 kg and the other has a mass of 75 kg, how must they adjust their speed before the collision to ensure that they will not drift across the ice while they are spinning? (Assume the ice is

frictionless.)

6. Two people are playing "catch" inside a closed railroad car that begins at rest. How would the car move as the ball is thrown back and forth? Why? (Assume that the car rolls with friction.)

7. A ball dropped from the third floor to the second floor of a house has a speed of 8 meters/s just before it hits the second floor. The same ball is caught on the second floor, and then dropped, from rest, to the first floor. What is its speed just before it hits the first floor? (Assume that the distance between floors is always the same.)

8. Explain the Coriolis force and how it affects a weather system moving north. The earth rotates from east to west. Which way do hurricanes rotate in the northern hemisphere? In the southern hemisphere? Explain with a sketch.

SECOND HOUR EXAM

Choose 5 questions. Show your thinking so that you can receive part credit when appropriate. Please write in blue or black ink. Good luck! You have 50 minutes, so do not spend too much time on any one of the questions. 15 points each.

1. The earth is closest to the sun in January. Is the earth moving in its orbit at its fastest speed or its slowest at this time? Why? Be explicit.

2. What is Newton's law of gravity? Give the equation and explain each term in it. Explain how you can tell from the equation that the law of gravity satisfies Newton's third law of motion.

3. How does Einstein's general theory of relativity describe gravity? Describe how it explains the bending of light by the sun. Describe the set of observations that verified this effect.

4. a) What is the difference between the situations when we can apply the special theory of relativity and the situations when we can apply the general theory of relativity?
b) Which case do we use to understand Einstein's elevator? Explain this example and its meaning.

5. Draw and describe a light clock inside a rocket, and how it would appear to an astronaut in the rocket and to someone who the rocket is passing by. Show how this leads to the effect of time dilation.

6. a) Sketch and distinguish between a transverse wave and a longitudinal wave.
b) Which is represented by a sound wave?
c) Sketch a wave and its second and third harmonics.

7. a) Describe the Doppler effect for sound and show with a sketch how it arises.
b) Describe and sketch how a shock wave arises. What is an example of a shock wave?

8. a) Sketch and describe constructive and destructive interference.
b) Show how a beam of coherent light, when directed through two adjacent slits, can interfere with itself to form light and dark bands on a screen. You will probably want to draw a sketch.

FINAL EXAM

Do 8 of the 10 questions. Show your thinking so that you can receive part credit whenever appropriate. Please write in blue or black ink and not in pencil. Please write in large letters on the upper right of the front cover of your bluebook which two questions you are omitting. Each question is worth 20 points; apportion your time appropriately.

1. a) Describe, with suitable sketches, how the Balmer series is produced by a hydrogen atom. (5 points)
b) At what frequency do we hear a train whistle if someone on the train hears the whistle at 440 Hz and the train is approaching us at a speed of 100 km/h? (The speed of sound is 330 m/s.) Show your work in detail. (10 points)
c) What is polarization? How do polarized sunglasses help cut glare? (5 points)

2. a) How does a laser work? (10 points)
b) List five current uses of lasers in research, business, or industry. (5 points)
c) Describe how to make a hologram. (5 points)

3. a) Illustrate the law of reflection, which relates the angle of incidence to the angle of reflection, for a plane mirror and for a curved telescope mirror. (5 points)
b) Explain why refraction takes place at the surface of a lens. (5 points)
c) What is total internal reflection, and what does it have to do with fiber optics? (5 points)
d) Explain a mirage, illustrating with a suitable sketch. (5 points)

4. a) Give a piece of evidence that light is a wave and a piece of evidence that light is a particle.
b) Describe Rutherford's scattering experiments and what they had to do with the structure of an atom.
c) Describe the assumptions that Bohr made in order to describe atomic spectra.
d) What is Heisenberg's uncertainty principle?

e) Describe the contribution of Schrodinger. (4 points each)

5. a) Sketch a nucleus of ordinary hydrogen, of deuterium, of ${}_6C^{12}$, and of helium. Use filled circles for protons and open circles for neutrons. (8 points)
b) What is the peninsula of stability in nuclei? (4 points)
c) Briefly define half-life. (4 points)
d) What role did Madame Curie play in the field of radioactivity? (4 points)

6. Discuss the pros and cons of fossil fuels, fission, and fusion for generating power for general use. (Be sure to organize this answer clearly, and not to use too much time on it.) (20 points)

7. a) Show by drawing +'s and -'s in suitable places what happens when you bring a positively charged rod near a metal ball.
b) When you bring two positively charged objects twice as close together as they were, by what factor and in what direction does the strength of the force change?
c) If you double the charge on one of the objects, the charge on the other object remains the same, and the distance between the objects remains the same, by what factor and in what direction does the force change?
d) How does a generator work? (5 points each)

8. a) Draw an electric monopole field. (3 points)
b) Draw an electric dipole field. (3 points)
c) Do both magnetic monopole and dipole fields exist? Explain. (4 points)
d) What is the significant feature of Maxwell's laws that led to the discovery of radio waves? (3 points)
e) Explain how natural magnetism is related to properties of the electrons inside the material. (4 points)
f) Sketch the earth's magnetic field. (3 points)

9. a) What are Kepler's laws of planetary motion?
b) What is Newton's law of gravity?
c) Explain the relation of Kepler's laws, Newton's law

of gravity, and finding the mass of the planet Jupiter.

d) How fast are divers moving when they hit the water after they fall from a height of 3 m? How fast are acrobats moving when they hit the net after they fall from a height of 10 m? The acceleration of gravity is about 10 m/s/s.

e) How long does it take the divers to hit the water and how long does it take the acrobats to hit the net under the above circumstances? (4 points each)

10. a) Describe the twin paradox and its resolution. Be sure that it is clear why it appears to be a paradox. (10 points)

b) Describe the quark theory, and how it explains the makeup of protons and neutrons (which are examples of hadrons) and mesons. (10 points)